高等职业教育通识类课程新形态教材

IT 职业英语
（第四版）

高巍巍　刘广敏　周　屹　刘　丽　编著

中国水利水电出版社
www.waterpub.com.cn
·北京·

内 容 提 要

本书以国际 IT 企业的实际应用规范为参考，汇集经验丰富的计算机及英语专业教师参与编写。全书以一名刚毕业的计算机专业学生进入公司工作的第一天以及在 IT 企业中接触到的各类常见技术场景为背景，侧重与工作密切结合的听、说、读、写多项实用英语技能，融合了信息技术和商务英语两大领域，培养学生流利使用商务口语、技术口语以及高效阅读技术文档的能力和写作商务 E-mail 和技术报告的能力。

本书具备行业性、实用性、交互性等多方面优势，旨在培养、提高学习者的 IT 英语应用能力，适用于高等院校计算机及相关专业、软件学院、各类职业院校和专业培训机构，也可作为专业人员自学英语、提高英语应用能力的参考书。

本书提供课件与听力文件，读者可以从中国水利水电出版社网站（www.waterpub.com.cn）或万水书苑网站（www.wsbookshow.com）免费下载。

图书在版编目（CIP）数据

IT职业英语 / 高巍巍等编著. -- 4版. -- 北京：中国水利水电出版社，2021.9
高等职业教育通识类课程新形态教材
ISBN 978-7-5170-9847-8

Ⅰ. ①I… Ⅱ. ①高… Ⅲ. ①IT产业－英语－高等学校－教材 Ⅳ. ①F49

中国版本图书馆CIP数据核字(2021)第165160号

策划编辑：石永峰　责任编辑：石永峰　加工编辑：黄卓群　封面设计：李 佳

书　名	高等职业教育通识类课程新形态教材 IT职业英语（第四版） IT ZHIYE YINGYU
作　者	高巍巍　刘广敏　周 屹　刘 丽　编著
出版发行	中国水利水电出版社 （北京市海淀区玉渊潭南路 1 号 D 座　100038） 网址：www.waterpub.com.cn E-mail：mchannel@263.net（万水） 　　　　sales@waterpub.com.cn 电话：（010）68367658（营销中心）、82562819（万水）
经　售	全国各地新华书店和相关出版物销售网点
排　版	北京万水电子信息有限公司
印　刷	三河市德贤弘印务有限公司
规　格	184mm×260mm　16 开本　13.75 印张　318 千字
版　次	2010 年 2 月第 1 版　2010 年 2 月第 1 次印刷 2021 年 9 月第 4 版　2021 年 9 月第 1 次印刷
印　数	0001—4000 册
定　价	39.00 元

凡购买我社图书，如有缺页、倒页、脱页的，本社营销中心负责调换

版权所有·侵权必究

第四版前言

为适应 IT 技术的国际化发展和计算机专业英语教学改革的要求，我们对《IT 职业英语》进行了修订再版。本次修订的编写目标为：紧跟 IT 技术的国际化发展节奏，既有反映最新 IT 技术的基本理论，又有贴近专业发展的实际应用。再版时主要对每个章节的技术阅读、快速阅读和补充阅读部分的内容进行了更新，并配有相应的习题帮助学习者掌握 IT 职业英语的学习方法，处理在工作中遇到的实际问题。读者可通过学习掌握最新的 IT 技术发展新知识，并具备参加各级各类 IT 英语考试的应试能力。

全书以一名初次进入 IT 企业的学生在工作中所接触的各类常见技术场景为背景，侧重与工作密切结合的听、说、读、写多项实用英语技能，融合了信息技术和商务英语两大领域，培养学生流利使用商务口语、技术口语以及高效阅读技术文档的能力和写作商务 E-mail、技术报告的能力。本书具备行业性、实用性、交互性等多方面优势，旨在培养、提高学习者的 IT 英语应用能力，适用于高等院校计算机及相关专业、软件学院、各类职业院校和专业培训机构，也可作为专业人员自学英语、提高英语应用能力的参考书。

全书共分两部分：

第一部分由 12 个单元组成，各单元结构如下：

- 阅读：包括技术阅读、快速阅读和辅助阅读，侧重培养学生对 IT 相关文章的阅读理解能力、增加学生的 IT 英语词汇量，其中辅助阅读部分为读者补充了大量 IT 文章，以扩展读者的视野。
- 听力：培养读者对英文对话的理解能力，增强读者在外企环境中的适应能力。
- 口语：侧重培养学生在企业中与上司、同事和客户的交流能力。
- 写作：主要介绍工作中经常使用的电子邮件及其他应用文的写作方法和技巧。侧重培养学生在特定情况下与人进行书面沟通的能力。

第二部分包括 12 个单元，主要是对会话和文章中经常使用的语法和句型进行综合讲解。

本书每个单元都附有大量练习题，帮助读者进行知识的巩固和提高，同时提供了听力部分的全文录音，扫描相应二维码即可收听。

本书由高巍巍、刘广敏、周屹、刘丽编著，其中第 1 单元和第 2 单元由高巍巍编写，第 3 单元至第 5 单元由刘广敏编写，第 6 单元由叶佳豪编写，第 7 单元至第 9 单元由周屹编写，第 10 单元由陈佳编写，第 11 单元和第 12 单元由刘丽编写；语法和写作部分由鞠鸿伟编写；听力部分由鞠鸿伟和黄文哲录音，全书由高巍巍统稿。

本书在再版过程中得到了中国水利水电出版社的大力支持，在此表示衷心感谢。同时，对在使用本书过程中反馈信息、提供修改意见的教师和读者表达特别的谢意。除封面署名作者外，参加本书编写和资料整理的还有王美佳、李晓峰、桑宇鹏、高炜、李放等，在此一并表示感谢。由于时间紧迫及编者水平有限，书中疏漏之处在所难免，恳请读者批评指正。

编者
2021 年 5 月

第一版前言

随着我国信息产业与国际的接轨,中国的 IT 行业在全球 IT 行业中占有越来越重要的地位,许多国外知名 IT 企业在中国投资,展开研发与生产活动。IT 行业的国际化发展使得企业对人才的需求向既懂得英语又懂技术的国际 IT 人才转变。本教程旨在提高计算机相关专业学生及 IT 行业从业人员的商务英语交流能力、技术阅读和商务写作能力,以国际 IT 企业的实际应用规范为编写参考,汇集了经验丰富的计算机与英语教师。

全书以一名刚毕业的计算机专业学生从进入公司工作的第一天以及整个业务活动为背景,侧重与工作密切结合的听、说、读、写多项实用英语技能,融合了信息技术和商务英语两大领域,培养学生流利地使用商务口语、技术口语以及高效阅读技术文档的能力和写作商务 E-mail、技术报告的能力。本教程适用于高等院校计算机及其相关专业、软件学院、各类职业院校和专业培训机构,也可作为个专业人员自学英语、提高英语应用能力的参考用书。

全书共分两大部分:

第一部分由 12 个单元组成,各单元结构如下:

- 阅读:包括技术阅读、快速阅读和辅助阅读。侧重于培养学生对 IT 相关文章的阅读理解能力,增加学生的 IT 英语词汇量。其中辅助阅读部分为读者补充了大量的 IT 文章,以扩充读者的视野。
- 听力:培养读者对英文对话的理解能力,增强读者在外企环境中的适应能力。
- 口语:侧重于培养学生在企业中与上司、同事和客户的交流能力。
- 写作:主要介绍在工作中经常使用的电子邮件以及其他应用文的写作方法和技巧。侧重培养学生在特定情况下与人进行书面沟通的能力。

第二部分为语法和句型部分,共包括 12 个单元,本部分主要是对会话和文章中经常使用的语法和句型进行综合讲解。

全书的每个单元都附有大量相关的练习题,帮助读者进行知识的巩固和提高。同时提供听力部分的全文录音,以方便广大读者的使用。

本书由高巍巍任主编,鞠鸿伟、黄玉妍、孙广丽任副主编,全书的语法和写作部分均由鞠鸿伟编写;除此之外,第 1、2 章由刘广敏编写,第 3、5、6 章由黄玉妍编写,第 4、12 章由孙广丽编写,第 7、8、9 章由高巍巍编写,第 10、11 章由杨巍巍编写。全书的听力部分由鞠鸿伟和黄文哲录音。

除封面署名和作者外,参加本书编写和整理资料的还有周洪玉、范晶、马宪敏、苍圣、陈丽、侯相茹、马玲、张丽明、张鑫瑜、高炜、李放等,在此一并表示感谢。由于时间紧迫,编者水平有限,疏漏在所难免,诚恳希望广大读者不吝指正。

编 者
2010 年 2 月

目　录

第四版前言

第一版前言

Unit 1　First Day at Work ················ 1
　Section One　Reading ················ 1
　Section Two　Listening ··············· 8
　Section Three　Speaking ·············· 8
　Section Four　Writing ··············· 10
　Transcript ······················ 16

Unit 2　Software ···················· 17
　Section One　Reading ··············· 17
　Section Two　Listening ·············· 22
　Section Three　Speaking ············· 23
　Section Four　Writing ··············· 24

Unit 3　Office Routine ················ 29
　Section One　Reading ··············· 29
　Section Two　Listening ·············· 37
　Section Three　Speaking ············· 38
　Section Four　Writing ··············· 38
　Transcript ······················ 44

Unit 4　Creative Software ·············· 45
　Section One　Reading ··············· 45
　Section Two　Listening ·············· 49
　Section Three　Speaking ············· 49
　Section Four　Writing ··············· 50
　Transcript ······················ 54

Unit 5　Communicate Online ············ 55
　Section One　Reading ··············· 55
　Section Two　Listening ·············· 61
　Section Three　Speaking ············· 63
　Section Four　Writing ··············· 63
　Transcript ······················ 67

Unit 6　Surf the Network ··············· 69
　Section One　Reading ··············· 69
　Section Two　Listening ·············· 76
　Section Three　Speaking ············· 77
　Section Four　Writing ··············· 77
　Transcript ······················ 83

Unit 7　Selling Products ··············· 85
　Section One　Reading ··············· 85
　Section Two　Listening ·············· 90
　Section Three　Speaking ············· 90
　Section Four　Writing ··············· 91
　Transcript ······················ 94

Unit 8　With Customers ··············· 95
　Section One　Reading ··············· 95
　Section Two　Listening ············· 102
　Section Three　Speaking ············ 102
　Section Four　Writing ·············· 103
　Transcript ····················· 105

Unit 9　Solutions ·················· 106
　Section One　Reading ·············· 106
　Section Two　Listening ············· 111
　Section Three　Speaking ············ 112
　Section Four　Writing ·············· 113
　Transcript ····················· 115

Unit 10　Computer Security ············ 116
　Section One　Reading ·············· 116
　Section Two　Listening ············· 121
　Section Three　Speaking ············ 122
　Section Four　Writing ·············· 122

  Transcript ·· 126
Unit 11 The Development Environment ········ 127
  Section One Reading ······················ 127
  Section Two Listening ····················· 135
  Section Three Speaking ··················· 135
  Section Four Writing ······················ 136
  Transcript ·· 145
Unit 12 New Technology ···························· 147
  Section One Reading ······················ 147
  Section Two Listening ····················· 153
  Section Three Speaking ··················· 153
  Section Four Writing ······················ 154
  Transcript ·· 156
Grammar ··· 158
  Unit 1 Tenses 时态 ······························ 158
  Unit 2 Passive Voice 被动语态 ················ 160

Unit 3 Sentences 句子 ····························· 161
Unit 4 Nominal Clauses 名词性从句 ··········· 164
Unit 5 Reported Speech 间接引语 ·············· 165
Unit 6 Adverbial Clauses 状语从句 ············· 166
Unit 7 Relative/Attributive Clauses
     定语从句 ···································· 168
Unit 8 Relative/Attributive Clauses and Appositive
     Clauses 定语从句和同位语从句 ········ 169
Unit 9 Modal Verbs 情态动词 ···················· 169
Unit 10 Inversion 倒装 ····························· 172
Unit 11 Non-finite Verbs 非谓语动词 ·········· 176
Unit 12 Infinitives&Gerunds 不定式&
      动名词 ···································· 179
Appendix A Glossary of Readings ··········· 181
Appendix B Key to Exercises ················· 189
Key to Grammar Exercises ·························· 210

Unit 1　First Day at Work

Section One　Reading

➢ Technical Reading

Working in an international IT company means that you have to deal with English all the time. Kevin needs to get familiar with the English equivalents of the technical terms he has learned. So the first class he needs to make up for is to learn the English name for each component of a computer.

Introduction to Computer Components

Computers are electronic machines which can accept data in a certain form, process the data and give the results of the processing in a specified format as information. Key **elements** in a computer system include the processor, **memory, input devices, output devices**, and **storage**. The input/output devices are sometimes called **peripheral** devices.

The processor is the **kernel** of any computer system, sometimes called the **Central Processing Unit** (CPU). In a way, it is the "brain" of the computer. It has three main sections: the control unit, the **Arithmetic and Logic Unit** (ALU) and the memory section.

Storage of data and software in a computer system is either temporary or permanent. **Random-Access Memory** (RAM), provides temporary storage of data and programs during processing within **chips**. Permanently installed and interchangeable disks provides permanent storage for data and programs for retrieval by the computer. In the early days, people used **floppy disk, hard disk**, CDs, DVDs, tapes, and **flash drives**. Now, with the development of technology, more and more people prefer to use mechanical hard disk, solid state hard disk and flash disk as storage media.

Computer systems use many devices for input purpose. Input devices enable data to go into the computer's memory. The most common input devices are the keyboard (for keyed input), a microphone (for voice and sound input), or a point-and-draw device, such as a mouse. The other typical input devices are **joy-stick**, digital camera and **scanner**.

Output devices enable us to extract the finished product from the system. For example, the computer shows the output on the **monitor** or prints the result on to paper by means of a **printer**. Output also can be routed to a video display or audio speakers.

On the rear panel of the computer there are several ports into which we can plug a wide range of peripherals—**modems, fax machines**, **optical drives** and scanners.

There are the main physical units of a computer system, generally known as the **configuration**.

Vocabulary

1. element — n. 元素
2. memory — n. 内存
3. input device — n. 输入设备
4. output device — n. 输出设备
5. storage — n. 存储，存储器
6. peripheral — n. 外围设备
7. kernel — n. 内核，核心
8. Central Processing Unit(CPU) — n. 中央处理器
9. Arithmetic and Logic Unit(ALU) — n. 算术逻辑单元
10. Random-Access Memory(RAM) — n. 随机存取存储器
11. chip — n. 芯片
12. floppy disk — n. 软盘
13. hard disk — n. 硬盘
14. flash drive — n. 闪存盘
15. joy-stick — n. 操纵杆
16. scanner — n. 扫描仪
17. monitor — n. 显示器
18. printer — n. 打印机
19. modem — n. 调制解调器
20. fax machine — n. 传真机
21. optical drive — n. 光盘驱动器
22. configuration — n. 配置

Exercise 1

True or False

1. (　) Key elements in a computer system include memory, input devices, and output devices.
2. (　) Keyboards, mice, input pens, and printers are input devices.

3. (　) The input/output devices are also called peripheral devices.
4. (　) CPU can be called the key element of a computer.
5. (　) The control section can be found in the CPU of a typical PC.

Exercise 2

There are 5 terms or phrases in the following box. Below the box are the explanations for these terms. Choose the correct explanation from a~e for each term by typing the corresponding letter.

1. software	_____
2. central processing unit	_____
3. peripheral devices	_____
4. hardware	_____
5. output	_____

a. the brain of the computer

b. programs which can be used on a particular computer system

c. results produced by a computer

d. physical parts that make up a computer system

e. hardware equipment attached to the computer

➢ Fast Reading

Text 1

The following is an advertisement with product specifications.

HP Z4 Workstation	Dell OptiPlex 7070MT	Dell Inspiron 5000 Urban 15	27-inch iMac 2020 verson
• Intel® Xeon® Scalable • Windows 10 Professional 64 • 8 GB 2666 MHz DDR4 ECC SDRAM (1×8 GB) • 8 DIMM • 1 TB 7200 rpm SATA • NVIDIA Quadro P2000 (5 GB) • Price: $2495.00	• 9th Generation Intel® Core™ i7 9700 processor (17M Cache, 3.0 GHz) • Windows 10 Home 64bit • 8GB DDR4 SDRAM at 2666MHz • 1TB 7200 HDD Hard Drive • Intel® HD Integrated Graphics • McAfee LiveSafe 12 Month Subscription • Price:$749.00	• 11th Generation Intel® Core™ i7-1165G7 processor (12M Cache, 2.8 GHz) • Windows 10, 64-bit, English • 15.6 inch LED Backlit Display with Truelife and FHD resolution (1920×1080) • 16GB DDR4 at 3200MHz • 512GB SSD Drive • Intel® HD Graphics • 1 Year Limited Warranty plus 1 Year Mail-In Service • 3.79 lbs • Price: $877.79	• Up to 3.7 GHz quad-core Intel Core i5 processor • Up to 32GB memory • 1TB or 3TB hard drive; 1TB or 3TB Fusion Drive; or up to 1TB flash storage1 • AMD Radeon Pro 570X, AMD Radeon Pro 580X • 27-inch (diagonal) LED-backlit widescreen display • 5120×2880 resolution • Price: $2521.00

Exercise 3

Read the descriptions of the four people and the four computers. Choose the most suitable computer for each person.

1. Steven is a accounting student. He needs a computer to write essays, assignments and letters.

2. Cindy is a manager of an advertising company. She needs a powerful system which will work with optical disks and multimedia applications, integrating text and picture with animation and voice annotations. Digitized images and sound occupy a lot of disk space.

3. Andy is a CAD engineer. His job involves computer-aided design, simulations and three-dimensional modeling. These applications require a lot of memory and a large drive.

4. Tanya is a sales representative. She needs a lightweight machine with which she can process orders and communicate with head office while she is on the road.

Text 2

The following is an introduction to two models of laptop.

Inspiron Mini 9

Inspiron Mini Laptops are designed to keep you connected. Our Minis or netbooks are also perfect for kids!

For those who live out of a suitcase and fly frequently, this ultra-portable provides integrated EV-DO.

- Connect with advanced wireless options
- Light and compact for an on-the-go lifestyle
- Dynamic & Customizable user interface

4 hours battery life & just 2.28 lbs

Inspiron Mini 9
Intel® Atom Processor® N270 (1.6GHz/533Mhz FSB/512K cache)

Operating System
Ubuntu Linux version 8.04.1

Display and Camera
Glossy 8.9 inch LED display (1024X600)

Inspiron 13

Inspiron laptops are our mainstream laptop brand, built for everyday use and available in colors and patterns. Give people something to talk about. Meet the new Inspiron™ 13 - the slim, brainy new 13.3" laptop from Dell.

- Up to Intel® Core™2 Duo processors
- Genuine Windows Vista® operating system
- Sleek design, Pacific Blue exterior with piano black accents

Inspiron 13
Intel® Core™ 2 Duo T6400 (2.0GHz/800Mhz FSB/2MB cache)

Operating System
Genuine Windows Vista® Home Basic Edition SP1

Display and Camera
Glossy, widescreen 13.3" LCD (1280x800) w/o Camera

Combo or DVD+RW Drive

Memory 512MB 2-DDR2 at 533MHz	8X Slot Load CD / DVD Burner (Dual Layer DVD+/-RW Drive)
Hard Drives 8GB Solid State Drive	Memory 3GB 2-Shared Dual Channel DDR2 (2 Dimms)
Video Card Intel Graphics Media Accelerator (GMA) 950	Hard Drives Size: 320GB 3-SATA Hard Drive (5400RPM)
Sound Base LCD Assembly	Video Card Intel Graphics Media Accelerator X3100
Wireless Networking Wireless 802.11g Mini Card	Wireless Networking Dell Wireless 1505 Wireless-N Mini-card
Primary Battery 39WHr Battery (4 cell)	Primary Battery 37Whr Lithium Ion Battery (4 cell)
System Color Option Alpine White	System Color Option Pacific Blue
Services 1Yr Ltd Warranty and Mail-In Service $1,928.00-$2,279.00	Services 1Yr Ltd Hardware Warranty, 4 InHome Service after Remote Diagnosis5 $1,728.00-$2,010.00

Exercise 4

Fill in the blanks according to the information from the text above.

1. If you are a businessman traveling around the world, you will choose model_____ for its following features:

   A. _____

   B. _____

2. If you are a software engineer, you will choose model_____ for its following features:

   A. _____

   B. _____

3. If you want to play games, you will choose _____.

4. If you travel frequently and you want to get a cheap laptop you will choose_____.

> Supplementary Reading

What Does a Scanner Do?

A **scanner** converts text or pictures into electronic codes that can be manipulated by the computer.

There are four different types of scanners: flatbed, sheet-fed, hand-held and drum. The most popular scanner used with a personal computer is the flatbed scanner.

In a flatbed scanner, the paper with the image is placed face down on a glass screen similar to a photocopier. Beneath the glass are the lighting and measurement devices. Once the scanner is activated, it reads the image as a series of **dots** and then generates a digitized image that is sent to the computer and stored as a file. The manufacture usually includes software which offers different ways of treating the scanned image.

A color scanner operates by using three rotating lamps, each of which has a different colored **filter**: red, green and blue. The resulting three separate images are combined into one by appropriate software.

To finish the scanning process, the new electronic image must be transferred to the computer. Scanners can be connected to computers in a variety of ways, such as USB, **FireWire** and Small Computer System Interface. It's also common to use a network that allows for a group of computers to connect to the same scanner. Special drivers that speak the scanner language of **TWAIN** are needed in order to read the output. These drivers sometimes come with application software such as **Photoshop** but usually come in the form of an installation CD with the scanner upon purchase. Once the image has been transferred to the computer, its information can be put into a number of different file types. **PDF** is the standard for document viewing because of its clarity, ease of use and **ubiquity**. A **PNG** is an uncompressed image that allows higher quality but uses more memory. A **JPG** is a compressed image that makes the file smaller, with lower image quality.

If the scanner is attached to a printer, it can simply scan the image and then make the desired amount of copies. If the scanner is by itself, it can be attached to a computer via a USB cord and the signal through the cord will send the scanned material to the desktop. If a high **resolution** printer is set up to receive the scanned document from the scanner or the computer, it will print off documents and items that were scanned just the same as if it were to be copied from a copy machine. Scanners can also scan in color so that the user can receive the added benefit of having professional, high-quality image results.

The following is an introduction to two models of scanner.

Color Scan XR from Sunrise

The Color Scan XR from Sunrise is a flatbed scanner with 600dpi of resolution and "9 *15" of scanning area.

Think of the possibilities.

You can enter data and graphic images directly into your applications-word processors or databases. You can get crisp, clean scans for color compositions, video and animation work.

It comes complete with its own image capture software which allows for color and grey retouching. and it's easy to use. What more could want for only £616? It couldn't be cheaper.

In the field of flatbeds, the Color Scan XR is a clear winner.

Color Scan XR

ScanPress 800

The ScanPress 800 is a self-calibrating, flated scanner with 800dpi of resolution. You can scan from black and white to 24-bit color. The package includes a hardware accelerator for JPEG compressing and decompression. JPEG technology saves disk space by compressing images up to 50 to 1.

In creating ScanPress 800, the manufacturers have chosen the highest technology to give you the best scans with the least effort. It produces images with high color definition and sharpness. And it comes with OCR software and Adobe Photoshop, so you can manipulate all the images you capture.

This is a fantastic machine you will love working with. And at only £1,037 it is an excellent investment.

Vocabulary

1. dot   n. 点
2. filter   n. 滤光器
3. FireWire   n. 火线
4. TWAIN   n. TWAIN 指 TWAIN 协议，是应用软件从计算机外设获取静态图像的国际标准
5. Photoshop   n. Adobe Photoshop，简称 PS，是一款由 Adobe Systems 开发和发行的图像处理软件
6. PDF   n. 便携文件格式，是由 Adobe 公司开发的独特的跨平台文件格式
7. ubiquity   n. 普遍性
8. PNG   n. 是一种位图文件（bitmap file）存储格式
9. JPG   n. JPG 全名是 JPEG。JPEG 图片以 24 位颜色存储单个光栅图像
10. resolution   n. 分辨率

Exercise 5

There are 5 terms or phrases in the following box. Below the box are the explanations for these terms. Choose the correct explanation from a~e for each term by typing the corresponding letter.

1. dpi	_____
2. "9×15"	_____
3. JPEG	_____
4. dot	_____
5. resolution	_____

a. Scanning area measured in inches

b. Dots per inch

c. A very small circular shape

d. The ability of the microscope or telescope to measure the angular separation of images that are close together

e. Joint Photographic Experts Group—a standard format in image compression. With the format your images can be compressed to $1/50^{th}$ of normal size, resulting in a substantial saving of disk space and time

Section Two Listening

Formal greeting and introductions.

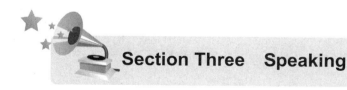

Section Three Speaking

Useful expression in office: small talk greeting and introduction.

Unit 1　First Day at Work

Exercise 6

假设销售部的经理吴敏和她的秘书一同拜访一个商务伙伴。首先吴敏需要做一个自我介绍，然后向商务伙伴介绍她的秘书。在这种场景下吴敏应该怎么说？

A: Allow me to _____. My name is Wu Min, a manager in the Sales Department.

B: _____, Miss Wu? Nice to meet you.

A: _____, here's my card.

B: Thank you. This is mine.

A: _____. This is Miss Li, my secretary.

B: Glad to meet you.

A: This is Mr. Chen.

C: Nice to meet you.

Exercise 7

假设你和研发部的同事王新在派对上遇见一个商务伙伴 Mr. King。你需要向商务伙伴介绍王新。在这种场景下你应该怎么说？

A: Wang Xin, _____ Mr. King, head of the England delegation, have you?

B: No, not yet.

A: Well, _____. Hello, Mr. King, I hope you're enjoying the party.

C: Yes, very much.

A: Mr. King, _____. Mr. Wang, from our R&D Department.

C: _____?

B: How do you do? I'm very glad to meet you, Mr. King.

Exercise 8

What would you say in the following situations?

1. Introduce yourselves in a less formal way.

2. Try to introduce two of your partners to each other.

3. Your boss says to you, "This is Mr. Smith. He is visiting us from England." What would you say to Mr. Smith?

4. You have been introduced to a client, but latter in the party you can not remember her name. What would you say?

5. The customer service manager, Mrs. Lee, doesn't know Cindy Morris, the new sales clerk. Would you introduce them to each other?

Section Four　Writing

Memos, notes and notices 备忘录、留言条和公告

A memo is used to communicate inside an organization, usually short but including the following parts（备忘录是一个机构内部沟通的工具，通常包括以下几个部分）:

1. To:　（致：）
2. From:　（由：）
3. Subject:　（事由：）
4. Date:　（日期：）
5. The actual message　（正文）

Memos can be less formal, formal or very formal.（备忘录文体可以是非正式的、正式的或非常正式的。）

For example:

Memo: Less formal（非正式文体）

Sample 1:

Memo	
To:	Office Managers
From:	Joe
Subject:	Room change for the next meeting
Date:	April 14th, 2020
The meeting on Friday, April 17th, has been changed to Room 302.	

Sample 2:

Memo	
To:	Everybody
From:	Anne
Subject:	Department meeting
Date:	July 14th, 2020
Our department meeting takes place at 1:30 in Room 402 every Friday afternoon.	

Sample 3:

Memo	
To:	Joe

From:	Ivy
Subject:	Routine check
Date:	August 18th, 2020
The top management will check all departments' routine work. Please get ready and cooperate well.	

Formal（正式文体）

Sample 1:

Memo	
To:	All teachers in our faculty
From:	Head of Faculty
Subject:	Training on the NIT exam
Date:	April 12th, 2020
Please tell your students that they can sign up for the NIT exam training course we are running from April 20th, 2020 to June 20th, 2020. Students will go to Mr. Zhang in Office 309 to get registered by next Friday April 17th, 2020. The training course costs 400 yuan. Lessons start at 6:00 pm and finish at 7:40 pm with 10 minutes break from 6:45 pm to 6:55 pm every Monday, Wednesday and Friday evening. Please contact me if there are any questions.	

Sample 2:

Memo	
To:	All Officers
From:	Human Resources Manager
Subject:	Training lesson
Date:	August 18th, 2020
There will be a training lesson for our customers on Wednesday July 25th to teach them how to use our after-sales service when they have problems with the software we developed for them. Please prepare what area each of you is responsible for and what problems you can solve for them. Thank you.	

Sample 3:

Memo	
To:	All employees
From:	Accounts office

Subject:	Lunch cards
Date:	Sep 15th, 2020

Everybody is arranged to eat in the canteen and supposed to use a new employee lunch card.

Please come to the accounts office to collect your card. Thank you.

Very Formal（非常正式文体）

Sample 1:

MEMORANDUM

To:	Z X Liu, General Manager
From:	Catherine Y L, Office Manager
Date:	12 April 2020
Subject:	Purchase of a Color Printer

1. Introduction

At the staff meeting on Wednesday, 9 April 2009, you asked for information about the possible purchase of a color printer. I would like to give these details.

2. Background

Since the eye-catchy posters are needed for advertising our products, staff has difficulty in printing colored pictures.

3. Advantages

Providing a color printer would help to make attractive posters and print colorful pictures.

4. Staff Opinion

Staff would like to use the color printer when necessary.

5. Cost

Details of suitable models are given below:

Brand	Model	Price
HP	Deskjet 5438(C9045D)	￥690
Epson	ME2	￥530
Lenovo	5510	￥500

6. Request

If this meets with your approval, we would appreciate it if you could authorize up to ￥700 for the purchase of the color printer.

Catherine Y L

Sample 2:

Memo	
To:	All Staff
From:	Accounts Manager
Subject:	NEW EXPENSES CLAIM SYSTEM
Date:	August 20th, 2020

> A corporate charge card and an automated expenses clam system will be used throughout the company.
>
> Employees who travel on company business will be sent a form by E-mail each month that shows all purchases made on the card.
>
> Employees indicate the business purpose of each purchase and then will be paid in cash. Line managers will monitor the claims and make a random check of the claims.

Sample 3:

Memo	
To:	Forever Estates
From:	Business Space
Subject:	Renting office space
Date:	June 20th, 2020

> Concerning the office space rent, the following is what we can offer you:
>
> 10% discount on the total price if you would like to rent our offices for 2 years;
>
> 4 offices left in the central plaza with no parking space but not in the Opera Place;
>
> Additional underground car park available near the central plaza.
>
> Please confirm the above and inform us about your final decision.

A note is used to keep your reader firmly in mind about matters or facts of a current situation. It should be simple and to the point. （留言条用来提示收条者当前事件的状况。它的特点是简单、直接。）

Notes include 3 parts, the salutation, the message and the writer's signature but very informal. （留言条由三部分组成：称呼、正文和署名，采用非正式文体。）

For example:

> Shelly,
> Here is your USB disk. Tell me you get it.
> Thanks.
> George

Henry,

　　Can't find the name list I need for the security meeting. If you've kept a record, please give another copy to me ASAP.

Thanks.

Ivy

Steve,

　　I took your mobile disk away for tomorrow because I need the document in it. I will return it quickly.

Thanks.

Ivy

　　A notice is an announcement containing information about an event. It should be quite short but very clear.

　　A notice needs a heading at the top, then the main announcement and at the bottom there should be the name of the writer and the date. （公告需要有标题、主要内容和结尾的署名、日期。）

　　For example:

LUNCH BREAK

　　From next Monday, lunch break time for Block 1 students will start from 11:35 and for Block 2 students it will begin from 11:45. Teachers can have their lunch break between 11:00 to 13:20.

Sue Miller, Personnel Manager
October 18th, 2019

A notice of Lost and Found

　　On the evening of March 4th, 2019, I found an MP3 in the English reading-room on the second floor of the new library in the east district of our university.

　　The MP3 can be generally described as follows. It is brand new and metallic gray in color. The portable MP3 is as big as a normal USB disk and as thin as a regular magazine.

　　The owner of the MP3 may contact me now. My mobile phone number is 138×××3544. Please make an appointment in advance.

Sincerely Yours,
Wang Yuan
March 22nd, 2019

> **Removal Notice**
>
> Our shop is now moved to a new business zone owing to the road construction. The current address is Apartment C, 88 Zhongshan Road. You are warmly welcome to the new site. Sorry for any inconvenience this may have caused you.
>
> Dream Tech Computers
> June 10th, 2019

The difference of memos, notes and notices（备忘录、留言条和公告的区别）

	Memos	Notes	Notices
1	to a group or an individual	to an individual	to anyone in particular
2	formal	informal	formal
3	use of formal job titles no use of: 1. abbreviations 2. ellipsis（省略、省略号） 3. contractions（缩写） 4. omissions of words	use of omissions of: 1. subjects 2. auxiliaries 3. articles	use of formal job titles no use of: 1. abbreviations 2. ellipsis 3. contractions 4. omissions of words

Exercise 9

1. *Write a memo*

You work for Acer Computer Company. Your Office Manager is Miss Xinyi Dong. She asks you to write a memo on behalf of her to all the company's representatives, informing them that the new supply of company-headed writing paper, notepads, and ballpens with the company's name and address, which are given to customers, has arrived and will be available to the representatives from next Monday from the office. She also needs to know the quantities of these items each representative requires.

2. *Write a note*

You are Amanda Ribera, the Human Resources Manager of the Blue Sky Company. You asked your colleague, Danny Brown, to interview a candidate for you because you will be away on a business trip next Tuesday. You have marked a few things you want to ask about the candidate. Your colleague agreed, so before you go, you need to write a note to thank him, also say when the interview takes place, what you've done to the candidate's C.V. and ask Danny to take notes for you.

3. *Write a notice*

You are Jay Yang, the General Affairs Manager of LTP Company. You have just informed by the electricity authorities that the electricity supply will be turned off tomorrow between 9am and

3pm. You can't afford to close the office for this long so you prepare some batteries for staff to use with some machines. Please write a notice to put on staff notice board in the reception area and apologize for any inconvenience.

Transcript

Listening 1

A: Good morning. My name is Kevin. I'm from Creative Software. How do you do? It's nice to meet you.

B: How do you do? I'm pleased to meet you, too. Welcome to IBM. I hope you enjoy your visit.

Listening 2

A: Mr. Lin, I'd like you to meet Amy Jin. She's our Human Resources Manager.

B: How do you do, Ms. Jin?

C: How do you do, Mr. Lin?

B: Nice to meet you.

C: It's nice to meet you, too.

Your turn: Try to introduce two of your partners to each other.

Unit 2 Software

Section One Reading

➢ Technical Reading

It all starts with your PC's operating system. To interact effectively with your operating system and any of the thousands of applications software packages on the market today, Kevin will need a working knowledge of its function and use.

Software Basic

Information provided by programs and data is known as **software**. Programs are sets of instructions that make the computer execute operations and tasks. There are two main types of software:

The **system software** refers to all the programs which control the basic functions of a computer. They include operating systems, system utilities (e.g. an anti-virus program, a back-up utility) and language translators (e.g. a complier—the software that translates instructions into machine code).

The **applications software** refers to all those applications—such as word processors and spreadsheets—which are used for specific purposes. Applications are usually stored on disks loaded into the RAM memory when activated by the user.

The **operating system** is the most important type of system software. It is usually supplied by the manufactures and comprises a set of programs and files that control the hardware and software resources of a computer system. It controls all the elements that the user sees, and it communicates directly with the computer. In most configuration, the OS is automatically loaded into the RAM section when the computer is started up. Popular operating systems include **Microsoft Windows family, Mac OS,** and **Linux**.

Utility software provides additional functionality to your operating system. This includes anti-virus software, crash-proof software, uninstaller software, and disk optimization software.

Within the realm of application software, you'll focus mostly on using personal productivity software:

- **Word processing and desktop publishing software** for creating mainly text documents.
- **Web authoring software** for building and maintaining a Web site.
- **Graphics software** for creating photos and art.
- **Presentation software** for building a slide presentation.
- **Spreadsheet software** for working with numbers, calculations, and graphs.
- **Personal information management and personal finance software** for managing personal information.
- **Communication software** for communicating with other people.

Vocabulary

1. software	*n.*	软件
2. system software	*n.*	系统软件
3. applications software	*n.*	应用软件
4. operating system	*n.*	操作系统
5. Microsoft Windows family	*n.*	微软家庭版视窗操作系统
6. Mac OS	*n.*	苹果操作系统
7. Linux	*n.*	一种开源操作系统
8. Utility software	*n.*	工具软件，通用软件
9. Word processing and desktop publishing software	*n.*	文字处理和桌面出版软件
10. Web authoring software	*n.*	网络编辑软件
11. Graphics software	*n.*	图形软件
12. Presentation software	*n.*	演示文稿软件
13. Spreadsheet software	*n.*	电子表格软件
14. Personal information management and personal finance software	*n.*	个人信息管理和财务软件
15. Communication software	*n.*	通信软件

Exercise 1

There are 10 terms or phrases in the following box. Below the box are the explanations for these terms. Choose the correct explanation from a–j for each term by typing the corresponding letter.

1. Operating system	_____
2. PDF	_____

3. Mac OS _____
4. Utility program _____
5. Platform _____
6. Bit _____
7. Password _____
8. Presentation software _____
9. Photo illustration software _____
10. Personal information management (PIM) system _____

a. The software that controls the execution of all applications and system software programs.
b. A word or phrase known only to the user. When entered, it permits the user to gain access to the system.
c. Portable document format.
d. Software application designed to help users organize random bits of information and to provide communications capabilities, such as E-mail and fax.
e. Software that enables the creation of original images and the modification of existing digitized images.
f. Software used to prepare information for multimedia presentations in meetings, reports, and oral presentations.
g. A definition of the standards by which software is developed and hardware is designed.
h. The operating system for the Apple family of microcomputers.
i. System software program that can assist with the day-to-day chores associated with computing and maintaining a computer system.
j. A binary digit (0 or 1).

➢ Fast Reading

Text 1

MS-DOS: This is the disk operating system developed in 1981 by Microsoft Corp. It is the standard OS for all IBM PC compatibles or clones. In this text-based operating system, you can communicate with the computer by typing commands that exist within its library. For example, some basic DOS commands include DIR (shows a list of all the files in a directory), COPY (makes a duplicate of a file), DEL (delete files).

Windows: A family of operating systems for personal computers. Windows dominates the personal computer world, running, by some estimates, on 90% of all personal computers. Like the Macintosh operating environment, Windows provides a graphical user interface (GUI), virtual memory management, multitasking, and support for many peripheral devices. In addition to Windows 7 and Windows 10, which run on Intel-based machines, Microsoft also sells Windows CE and Windows NT family that run on a variety of hardware platforms.

UNIX: UNIX is a multiuser, multitasking operating system that is widely used as ht master control program in workstations and especially servers. This operating system designed by Bell Laboratories in the USA for minicomputers, has been widely adopted by many corporate installations. It is written in C language. It has become an operating environment for software development, available for any type of machine, from IBM to Macs to Cray supercomputers. UNIX is the most commonly used system for advanced CAD programs.

Linux: A UNIX-like OS kernel that runs on a variety of hardware platforms including Intel, Alpha and Sun SPARC, etc. Linux is open source software, which is freely available.

Mac OS: Mac system is a special system of Macintosh (Apple), it is a graphical operating system based on Unix kernel which is Developed by Apple Corp. The Mac operating system to OS 10, code named MAC OS X (Rome digital writing X 10). The new system is very reliable, many features and services embodies the idea of Apple Corp.

Java OS: This is designed to execute Java programs on Web-based PCs. It's written in Java, a programming language that allows Web pages to display animation, play music, etc. The central component of Java OS is known as the Java Virtual Machine.

Exercise 2

Answer the questions:

1. What dose "MS-DOS" stand for?
2. What is the basic DOS command for copying a file?
3. What is the abbreviation for "International Business Machines"?
4. Which company developed UNIX?

Text 2

Linux is an operating system that was initially created as a hobby by a young student, Linus Torvalds, at the University of Helsinki in Finland. Linus had an interest in Minix, a small UNIX system, and decided to develop a system that exceeded the Minix standards. He began his work in 1991 when he released version 0.02 and worked steadily until 1994 when version 1.0 of the Linux Kernel was released. The kernel, at the heart of all Linux systems, is developed and released under the GNU General Public License and its source code is freely available to everyone. It is this kernel that forms the base around which a Linux operating system is developed. There are now literally hundreds of companies and organizations and an equal number of individuals that have released their own versions of operating systems based on the Linux kernel.

Apart from the fact that it's freely distributed, Linux's functionality, adaptability and robustness, have made it the main alternative for proprietary UNIX and Microsoft operating systems. IBM, Hewlett-Packard and other giants of the computing world have embraced Linux and support its ongoing development. More than a decade after its initial release, Linux is being adopted worldwide primarily as a server platform. Its use as a home and office desktop operating system is also on the rise. The operating system can also be incorporated directly into microchips in a process called "embedding" and is increasingly being used this way in appliances and devices.

Throughout most of the 1990's, tech pundits, largely unaware of Linux's potential, dismissed it as a computer hobbyist project, unsuitable for the general public's computing needs. Through the efforts of developers of desktop management systems such as KDE and GNOME, office suite project OpenOffice.org and the Mozilla web browser project, to name only a few, there are now a wide range of applications that run on Linux and it can be used by anyone regardless of his/her knowledge of computers.

Exercise 3

1. About the starting of Linux, which of the following is true?
   A. It was developed by a group of software engineers.
   B. It was initiated by a US student.
   C. Microsoft experts initiated this language.
   D. A Finland student started Linux.

2. The meaning of word "kernel" is closest to which of the following?
   A. Peripheral.   B. Shell.
   C. Core.   D. Bottle.

3. Linux is welcomed for all the following reasons EXCEPT_____.
   A. free distribution   B. adaptability
   C. user friendliness   D. robustness

4. What can be inferred about Linux in the 1990's?
   A. It was very popular.
   B. It was not very popular.
   C. It was expensive.
   D. It was not developed yet.

5. About the future of Linux, which of the following is right?
   A. It will remain active only as a server operating system.
   B. It will totally replace Windows systems.
   C. It will be increasingly used by home users.
   D. It's hard to tell.

➢ Supplementary Reading

Basic DOS Commands

Through the 1980s, the most popular microcomputer operating system was MS-DOS. The MS is short for Microsoft and DOS is an abbreviation for disk operating system, meaning that ist is loaded from disk. MS-DOS was strictly text-based, command-driven software. That is, we issued commands directly to DOS(the MS-DOS nickname)by entering them on the keyboard, one character at a time. For example, if you wished issue a command to copy a word processing document from one disk to another for your friend, you might have entered "copy c:\myfile.text a:\yourfile.txt"via the keyboard at the DOS prompt "c:\>".

C:\>copy c:\myfile.txt a:\yourfile.txt

Command-driven DOS, in particular ,demand strict adherence to command syntax, which are the rules for entering commands, such as word spacing, punctuation, and so on.

Exercise 4

Match the DOS commands on the left with the explanations on the right. Some commands are abbreviations of English words.

1. FORMAT	A. erases files and programs from your disk
2. CD (or CHDIR)	B. copies all files from one floppy disk to another
3. DIR	C. changes your current directory
4. MD (or MKDIR)	D. initializes a floppy disk and prepares it for use
5. DISKCOPY	E. displays a list of the files of a disk or directory
6. BACKUP	F. changes names of your files
7. REN (or RENAME)	G. creates a subdirectory
8. DEL	H. A copy of a program or data file on a floppy disk, tape, CD, or hard drive

Section Two Listening

Short Conversations: *In this section you will hear 5 short conversations. After each conversation, you will hear 1 question and choose the best answer from the 4 alternatives marked*

A, B, C and D. You will hear each conversation and question ONLY ONCE.

1.
   A. The woman is willing to help the man.
   B. The woman doesn't have any time.
   C. The woman doesn't know what a report is.
   D. The woman wants to leave her office.
2.
   A. Go back to Ace Computer.
   B. Meet with the woman.
   C. Wait for Mr. Black.
   D. Find the way to the Meeting Room.
3.
   A. A written report.
   B. Financial support.
   C. Some extra information.
   D. His approval to a report.
4.
   A. He doesn't understand it either.
   B. He has lost his way to the office.
   C. He didn't really notice the meeting.
   D. He can't see the writing on the board.
5.
   A. The woman can get it done easily.
   B. The woman should be very careful.
   C. The man should help hand by hand.
   D. The man should stop talking.

Section Three Speaking

Exercise 5

What would you say in the following situations?

1. Please introduce your technical background.
2. What have you learnt about software programming?

3. A customer is calling to speak with your colleague, Paul, but right now he is on another call.

   a. Ask the customer his name, and if he is calling about anything urgent.

   b. Politely ask him to hold for a while.

4. You are new in your company. You want to go to the financial department but you have lost your way.

   a. Ask a colleague how to go there.

   b. You didn't hear what he said, so politely ask him to repeat.

   c. Thank him for his help.

5. You are the receptionist of your company. A new customer has just arrived.

   a. Politely ask him if he is Mr. Smith.

   b. Politely ask him if he needs some coffee.

   c. Reply politely to his thanks.

Section Four Writing

Filling the forms 填表

In our daily life, we are asked to fill in many kinds of forms such as application forms, order forms, staff appraisal forms, etc.（在日常生活中，我们需要填写各种表格，如申请表、订货表、员工评估表等。）

The following personal information is often needed (sample answers are given):（经常需要填写下列个人信息，后面是参考答案。）

Title 称呼/头衔	Mr./Mrs./Miss./Ms. 先生/夫人/小姐/女士
First name/Given name 名字	Jianguo/Lihua 建国/丽华
Surname/Last name/Family name 姓	Li/Wang 李/王
Name/Full name 全名	Li Jianguo/Wang Lihua 李建国/王丽华
Sex/Gender 性别	Male/Female 男/女
Date of Birth 出生日期	12th April, 1976/8 November, 1978 1976 年 4 月 12 日/1978 年 11 月 8 日
Place of Birth 出生地或籍贯	Harbin/Dalian 哈尔滨/大连
Country of origin 国籍	China 中国
Age 年龄	33/31 33 岁/31 岁
Height 身高	1.80m/1.68m 1 米 80/1 米 68

Unit 2 Software

Weight 体重	90kg/55kg　90 公斤/55 公斤
Marital Status 婚姻状况	Married/Single/Divorced/Widowed 已婚/未婚/离异/鳏/寡
No. of dependent children 抚养子女数量	1　1个
Employment status 就业状况	Employed full-time/Employed part-time/Self-employed/Unemployed/Retired　全职/兼职/自行创业/失业/退休
Current address 现住址	××Xiqiao St, Daoli District, Harbin, Heilongjiang Province, China　中国黑龙江省哈尔滨市道里区西桥街××号
Postcode 邮编	150001　（150001）
Phone Number 电话(Home)（宅）0451-2289××××	(Office)（办）：0451-8812×××× ／ (Mobile)（手机）：1380×××879
E-mail (address)　电子信箱:marywang@yahoo.co.uk	
ID Number 身份证号码:230101×××××××7339	
Family members'(parents or spouse) name/age 家庭成员（父母或配偶）姓名/年龄	Husband/wife; daughters/sons, etc. 丈夫/妻子；女儿/儿子等
Certificates/Qualifications 证书/资格证书	MCP(Microsoft Certified Professional) MCSE(Microsoft Certified Systems Engineer) MCDBA(Microsoft Certified Database Administrator) SCJP(Sun Certified JAVA Programmer)/SCJD(Sun Certified JAVA Developer) EPTIP(English Proficiency Test for IT Professionals)
Qualifications 资格证书	Diploma　毕业证书/Associate diploma 大专毕业证书/Undergraduate diploma 本科毕业证书/Postgraduate diploma 研究生毕业证书
Degrees 学位	Bachelor's Degree 学士学位/Master's Degree 硕士学位/Doctor's Degree 博士学位
Major 专业	computer science 计算机科学
School 学校	Harbin Institute of Technology 哈尔滨工业大学/Heilongjiang University 黑龙江大学/Harbin Normal University 哈尔滨师范大学
Job title/Position/Occupation 职位	Software design engineer 软件设计师
Department 部门	Personnel 人事部/Human Resources 人力资源部/Financial Department 财务部/Accounts 会计部/Purchasing 采购部/Production 生产部/Marketing 市场营销部/Sales 销售部/Public Relations 公关部/Advertising 广告部/Dispatch 运输部/After-sales 售后服务部/Maintenance 维修部/Supplies 供应部或物资部/Planning 规划部或策划部/Distribution 发行部或物流部/Quality control 质检部/Research and Development 研发部
Company name 公司名称	Microsoft Corp.微软公司
Income Details 收入详情	annual income ￥100,000　年收入人民币 100,000 元

For example:

1. Job Application Form　工作申请表

Dream Software Co. Ltd Job Application Form	
Position applied for	a software engineer on software design and development
Title: (Mr./Mrs./Ms./Miss)	Mr.
Full name:	Li Junfeng
Nationality:	China
Marital Status:	Single
Date of Birth:	August 25, 1982
Address:	228 Xidazhi St, Nangang District, Harbin, China
Phone Number:	0451-84840242
E-mail:	lijunfengxsh@freemail.yeah.net
Current employment:	Systems Analyst, Liancheng Information Technology Co. Ltd(2006-2008)
Higher education:	B.Sc., Computer Science, Harbin Institute of Technology (2001-2005)
Professional qualifications:	Bachelor's of Science Degree in Computer Science
Computer skills:	Applications: Microsoft Office; Programming: C, C++, Visual Basic
Language skills:	English
Interests:	Reading, Writing, Music, Travelling
Signature:	*Li Junfeng*
Date:	June 18, 2008

2. Printer Order Form 打印机订货表

PRODUCT DETAILS			
New Color Laser Printer	Model		
Specification	560Pro	Tek200	Pro-jet
Width	50cm	40cm	62cm
Depth	53cm	50cm	49cm
Height	40cm	33cm	37cm
Pages per min.	3	6	4-5
Memory	12MB	24MB	20MB
Price	￥299	￥349	￥390

3. Staff Appraisal Form 员工评价表

Dragon Tech Company Staff Performance Appraisal Form	
Name	Sun Lin
Position	Software Developer
Department	System Software Dept
Date	30 May, 2008
Question	**Answer**
What do you consider the most important responsibilities of your job?	Pass the software test
What has gone well in your job?	Write the functions
What has helped you carry out your work successfully?	The previous samples
What causes you concern or frustration in your job?	Be creative
What do you think has been difficult in your job?	Work overtime
What improvements do you think you have made in the way you do your job?	Realise the function designed
Appraiser	Chen Yishuai
Appraisal Scope	**Comments**
Progress made to achieve agreed objectives	needs effort
Teamwork and Co-operation including sharing information and knowledge	good
Communication Skills (ability to give and receive colleagues' information, willingness to listen to and understand colleagues)	good
Job Knowledge and Competence (capability of doing the current job)	good
Potentiality	very potential

Exercise 6

Fill in the form. It is an application form to open a bank account.

CITY BANK ACCOUNT APPLICATION FORM	
Surname	
First name	
Gender	
Date of birth	
Country of Origin	

Present address			
Post code			
When did you move to this address?			
Permanent address (if different from above)			
Telephone no. (home)			
Telephone no. (office)			
Marital Status			
No. of dependent children			
Residential details (owned or rented)			
Employment status			
Income details (per year)			
Signature		Date	

Unit 3 Office Routine

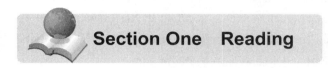

Section One Reading

➢ Technical Reading

Nowadays, when you use computer, you may interact with it through GUIs. The following text shows you what is GUI, and how many types of GUIs can people use today.

Graphical User Interfaces

When you first turn on your computer, you'll see a screen similar to that in Figure 3.1. This is called your desktop, which is a **graphical user interface**. A graphical user interface (GUI) is a graphic or icon-driven interface on which your use your mouse to start software (such as a Web browser), use that software, and initiate various other functions.

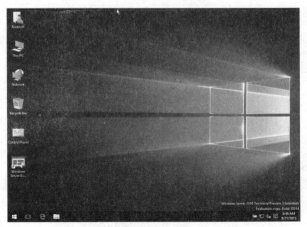

Figure 3.1

A graphical user interface (GUI) uses graphics, icons and pointers to make the computer more user friendly. The command required can be specified by clicking or double-clicking with a mouse. The menu screen or desktop can easily be edited to include extra options.

The GUI was first developed for the Apple **Macintosh** using WIMPs (Windows, Icons, Mice and Pointers). GUIs make the computer much more user friendly and more suited to the casual IT user.

Any GUI typically contains both buttons and icons. A button is a graphic representation of something that you click on once with the left mouse button. An icon is a graphic representation of something that you click on twice or double-click. You can start Internet Explorer (one of many available web browsers) with either the button or the icon. So, you can click once on the Internet Explorer button located at the bottom on the screen (the area is called the task bar). Or you can double click on the Internet Explorer icon located in the middle left of the screen. No matter which you choose, Internet Explorer will start and display a web site for you.

A good user interface is important because when you buy a program you want to use it easily. Moreover, a graphical user interface saves a lot of time: you don't need to memorize commands in order to execute an application; you only have to point and click so that its content appears on the screen.

- Features of GUIs

GUIs such as Windows include many features that make dialogue between the user and computer easier. These include:

- PULL DOWN MENUS: A menu that the user "pulls down" from a name in the menu bar at the top of the screen by selecting the name with mouse.
- ICONS: An **icon** is a small picture on the screen, used as a short cut to a function such as open a file, save, or print. Nearly all Windows software uses icons, which often can be customized to suit the user. Different programs often use the same icons for common functions.
- FOLDERS: Containers for documents and applications, similar to the subdirectories of a PC platform.
- DIALOGUE BOXES: Many Windows programs include dialogue boxes on the screen and asking the user certain questions. This is very easy for the user and makes sure that all the questions that need to be answered are answered. A dialogue box may be part of a wizard – a quick way of automatically performing a task in a program.
- FORMS: An on-screen form looks like a form on a piece of paper and has boxes to be filled in. It is easy for a novice user to enter information. List boxes can be used to restrict information entered to a list of **alternatives**.
- WYSIWYG: Most Windows programs, particularly word-processing and DTP programs are WYSIWYG (What You See Is What You Get). What you see on the screen is what is printed on the paper.

Vocabulary

1. graphical user interface   n. 图形用户界面
2. Macintosh   n. 麦金塔电脑，Apple 公司于 1984 年推出的一种微机
3. icon   n. 图标
4. alternative   n. 二中选一，可供选择的事物

Exercise 1

Read the article and find answers to these questions.

1. What dose the abbreviation "GUI" stand for?
2. What is the contribution of GUIs to the development of graphic environments?
3. What dose the acronym "WIMP" mean?

➢ Fast Reading

Text 1

When you come across a new software, you may need an easy-to-read instruction to help you. A tutorial can serve this purpose. The following tutorial shows you how to use Microsoft Office, a most frequently used program in work.

A Tutorial on Office Software Suit

Microsoft Office is a software suite that consists of different **applications** that complete different activities. MS Office is by far the most widely recognized software suite in the world.

Microsoft Word: Microsoft Word **provides** powerful tools for creating and sharing professional word processing documents.

Microsoft Excel: With Microsoft Excel, you can create detailed **spreadsheets** for viewing and **collaboration**. Create **customized formulas** for your data and **analyze** it with the easy to construct **charts**.

Microsoft PowerPoint: Microsoft PowerPoint provides a complete set of tools for creating powerful **presentations**. Organize and **format** your material easily, **illustrate** your points with your own images or **clip art**, and even broadcast your presentations over the Web.

Microsoft Access: Microsoft Access gives you powerful new tools for managing your **databases**. Share your database with **co-workers** over a **network**, find and retrieve information quickly, and **take advantage of** automated, pre-packaged wizards and solutions to quickly create databases.

Microsoft Publisher: Microsoft Publisher helps you easily create, customize, and **publish** materials such as **newsletters**, brochures, flyers, **catalogs**, and **Web sites**. Publish easily on your **desktop** printer.

Vocabulary

1. application   *n.* 应用，应用软件
2. provide   *vt.& vi.* 提供，供给，供应
3. spreadsheet   *n.* 电子数据表
4. collaboration   *n.* 合作，协作
5. customize   *vt.* 定制，定做
6. formula   *n.* 准则，原则，公式
7. analyze   *vt.* 分析，分解，解释
8. chart   *n.* 图表
9. presentation   *n.* 提供，显示，外观，报告，表演
10. format   *vt.* 使格式化，编排格式
                           n. 设计，安排，格式，样式，版式
11. illustrate   *vt.* 给……加插图，说明，阐明，表明
12. clip art   *n.* 剪贴画
13. database   *n.* 数据库，资料库
14. co-worker   *n.* 共同工作者，合作者，同事，帮手
15. network   *n.* 网络
16. take advantage of   *v.* 利用
17. publish   *vt.&vi.* 出版，公布
18. newsletter   *n.* 通讯，简报
19. catalog   *n.* 目录，目录册
20. Web site   *n.* 网站
21. desktop   *n.* 桌面

Exercise 2

True or False

1. (   ) Microsoft Office is a software suite that consists of four applications that complete different activities.
2. (   ) Microsoft Access provides powerful tools for creating and sharing professional word processing documents.
3. (   ) Microsoft Access gives you powerful new tools for managing your databases.
4. (   ) Microsoft Word helps you easily create, customize, and publish materials.
5. (   ) Microsoft PowerPoint can broadcast your presentations over the Web.

Unit 3 Office Routine

Text 2
Microsoft Word Processing Features Overview

This word processing document illustrates features common to most word processing software. You can create special effects with the drawing tool and border features. The watermark feature lets you add a drawing, a company logo, headline sized text (such as the "PRIOITY" in this example), or any image behind the printed document text. In the electronic world, documents are "networked" with hyperlinks (references to different sections of an electronic document or to other related electronic documents on the local computer or on the Internet). Even the callouts, which label the features, are a word processing feature. Not shown is the editing feature that lets you add editorial remarks and make corrections to an original document. This feature is helpful when several people review a document prior to publication.

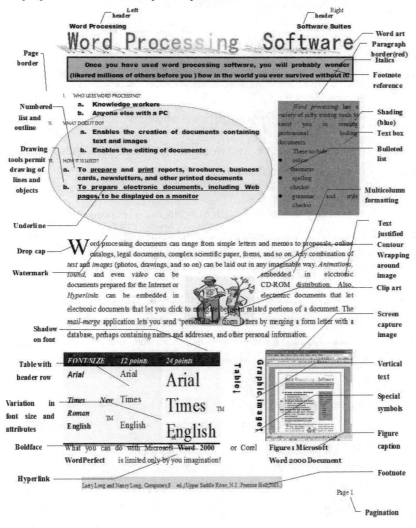

Figure 3.2

Exercise 3

True or False

1. (　) From the text above we can know creating Word art using Microsoft Word is pretty simple.
2. (　) You can not create special effects such as watermark using Microsoft Word.
3. (　) Hyperlink in Microsoft Word references to different sections of an electronic document or to other related electronic documents on the local computer or on the Internet.
4. (　) Images which are inserted as watermark are not different from images which are inserted as files of images in Microsoft Word.
5. (　) Special symbols and math formula can not be used in Microsoft Word.

➢ Supplementary Reading

Word Processing Software

Using a computer to create, edit, and print **documents**. Of all computer **applications**, word processing is the most common. To **perform** word processing, you need a computer, a special program called a word processor, and a **printer**. A word processor enables you to create a document, store it electronically on a **disk, display** it on a **screen**, modify it by entering **commands** and **characters** from the **keyboard**, and print it on a printer.

The great advantage of word processing over using a **typewriter** is that you can make changes without retyping the **entire** document. If you make a typing mistake, you simply back up the **cursor** and **correct** your mistake. If you want to delete a **paragraph**, you simply remove it, without leaving a trace. It is equally easy to insert a word, sentence, or paragraph in the middle of a document. Word processors also make it easy to move sections of text from one place to another within a document, or between documents. When you have made all the changes you want, you can send the file to a printer to get a **hardcopy**.

Word processing is more than just typing, however. Features such as **search** and **replace** allow users to find a particular phrase or word no matter where it is in a body of text. This becomes more useful as the amount of text grows.

Word processor usually include different ways to view the text. Some include a view that displays the text with editor's marks that show hidden characters or commands (spaces, returns, paragraph endings, applied styles, etc.). Many word processors include the ability to show exactly how the text will appear on paper when printed. This is called WYSIWYG (what you see is what you get). WYSIWYG shows **bold**, *italic*, underline and other type style characteristics on the screen so that the user can clearly see what he or she is typing. Another feature is the correct display of different typefaces and **format** characteristics (margins, indents, super and sub-scripted characters, etc.). This allows the user to plan the document more accurately and reduces the frustration of printing something that doesn't look right.

Many word processors now have so many features that they approach the capabilities of **layout applications** for desktop publishing. They can import graphics, formant multiple columns of text, run text around graphics, etc.

Two important features offered by word processors are **automatic hyphenation** and **mail merging**. Automatic hyphenation is the splitting of a word between two lines so that the text will fit better on the page. The word processor constantly monitors words typed and when it reaches the end of a line, if a word is too long to fit, it checks that word in a hyphenation dictionary. This dictionary contains a list of words with the preferred places to split it. If one of these cases fits part processor at the end of the line, the word processor splits the word, adds a hyphen at the happens extremely fast and gives text a more polished and professional look.

Mail merge applications are largely responsible for the explosion of "personalized" mail. Form letters with designated spaces for names and addresses are stored as documents with links to lists of names and addresses of potential buyers or clients. By designating what information goes into which blank space, a computer can process a huge amount of correspondence substituting the "personal" information into a form letter. The final document appears to be typed specifically to the person addressed.

Many word processors can also generate tables of numbers or figures, sophisticated indexes and comprehensive tables of contents.

Vocabulary

1.document		n. 公文，文件，文献
2.application		n. 应用软件
3.perform		vt.&vi. 执行，表演，扮演
4. printer		n. 打印机
5.disk		n. 磁盘，光盘
6.display		vt. 列，显示，展览
7.screen		n. 显示器
8.command		n. 命令，指令
9.character		n. 字符
10.keyboard		n. 键盘
11.typewriter		n. 打字机
12.entire		adj. 全部的，整体的
13.cursor		n. 光标

14.correct	*adj.* 正确的
15.paragraph	*n.* 段落
16.hardcopy	*n.* 硬拷贝
17.search	*vt.&vi.* 查找
18. replace	*vt.* 替换
19.bold	*n.* 粗体
20.format	*n.* 格式
21.layout application	*n.* 布局应用程序
22. automatic hyphenation	*n.* 自动断字
23.mail merging	*n.* 邮件合并

Exercise 4

A. *Look at the words in the box and complete the following sentences with them. Use the information in the text or Glossary if necessary.*

type style	WYSIWYG	format	indent
Font menu	Justification	Mail merging	

1. _____ stands for 'what you see is what you get'. It means that your printout will precisely match what you see on the screen.
2. _____ refers to the process by which the space between the words in a line is divided evenly to make the text flush with both left and right margins.
3. You can change font by selecting the font name and point size form the _____.
4. _____ refers to a distinguishing visual characteristic of a typeface; "italic", for example is a _____ that may be used with a number of typefaces.
5. The_____ menu of a word processor allows you to set margins, page numbers, spaces between columns and paragraph justifications.
6. _____ enables you to combine two files, one containing names and addresses and the other containing a standard letter.
7. An _____ is the distance between the beginning of a line and the left margin, or the end of a line and the right margin. Indented text is usually narrower than text without _____.

B. *There are 6 terms or phrases in the following box. Below the box are the explanations for these terms. Choose the correct explanation from a – f for each term by typing the corresponding letter.*

Unit 3　Office Routine

1. retrieve	_____
2. typefaces	_____
3. header	_____
4. footer	_____
5. subscripted character	_____
6. hyphenation	_____

a. text printed in the top margin

b. recover information from a computer system

c. letter, number or symbol that appears below the baseline of the row of type; commonly used in the math's formulas

d. text printed in the bottom margin

e. division of words into syllables by the short dash or hyphen

f. style for a set of characters; sometimes called "fonts"

Section Two　Listening

Listening 1

Transfer calling

总经理办公室秘书张伟接到客户王强先生的电话，他希望同总经理刘梅对话，然而刘梅不在公司，在这种情况下，他们会说些什么？

Listening 2

Leaving message

Emily 给在电脑公司工作的朋友 Kate 打电话，安排一个晚会，但是 Kate 现在正在接别的电话，在这种情况下，她们将说些什么？

Exercise 5

Listen and fill in the following outline.

Notes outline

1. The first speaker is the _____ of Mr. Lin's office.

2. Mr. White wants to _____ with Mr. Lin about the _____.
3. In fact, Mr. White wants to speak to _____.
4. Mr. Patrick Lin is _____.
5. The secretary will _____ Mr. White's telephone to the President's office.

Section Three Speaking

Useful expression of making telephone calls

Exercise 6

Kevin is inviting his colleague Peter to watch a football match with him. Peter really loves football games, but he has another engagement. What would they say in this situation?

Kevin: Good morning, ____①____! How are you today?

Peter: ____②____, Kevin!____③____. Thank you, ____④____?

Kevin: I'm fine too. Do you like watching football games?

Peter: ____⑤____, I'm a big football fan. ____⑥____?

Kevin: Then would you like to go to a football game with me tonight? It's a final game of CFA.

Peter: ____⑦____, I'd love to. But ____⑧____.

Kevin: That's too bad. What a shame. Perhaps next time.

Section Four Writing

Business Letter 商务信函

A business letter is a formal means of communication between two people, a person and a corporation, or two corporations. There are five parts of a business letter. （商务信函是个人之间、个人与公司之间以及公司与公司之间比较正式的沟通工具。它由五部分组成。）

1. Addresses and date （地址和日期）

Writer's (company) address: top right hand corner

Date: top right hand corner below the writer's address

Receiver's address: beneath date on left hand side (Make it as complete as possible; include titles and names if you know them.)

2. Greeting（称呼）

Correct Salutation: (Salutation is also called the greeting. The greeting in a business letter is always formal.)

"Dear (title + surname)," – if you know the person's gender, marital status and surname

"Dear Sir/Madam," – if you don't know the person's name or if you are writing to a company or organization

3. Body（正文）

The body of the letter begins below the salutation and uses clear paragraphs. State the purpose in the first opening paragraph. Describe what's wanted in the middle of the letter's body, and request specific action at the end of the body.

4. Complimentary close（结尾）

Complimentary close: Yours sincerely after "Dear (title + surname),"

Yours faithfully after "Dear Sir/Madam,"

5. Signature（署名）

Sign the writer's name and write the position under the name

Example of layout 商务信函格式和样本

	26 Hongtu Street
	Daowai District
	Harbin
	9 September, 2019
Mr. Xia Fengyi	
82 Railway Road	
Heihe	
Heilongjiang Province	
Dear Mr. Xia,	
....	
Yours sincerely,	
Yan Li	
Sales Manager	

THE 7 C's OF WRITING BUSINESS LETTERS 商务英语写作的7个注意事项

Clear 清晰: Express yourself clearly to make sure to use simple, suitable language.

Concise 简洁: Keep your sentences short, to the point.

Complete 完整: Include all necessary facts or background information to support you.

Correct 正确: Use specific facts or figures; Give specific time with date; etc.

Courteous 礼貌: Be polite and use formal language.

Considerate 周到: Keep your reader's needs in mind as you write.

Convincing 可信: Present yourself of reliability and competence. Reinforce and make yourself more believable.

Useful functional language 有用的功能性语言

- Stating the reason for writing 阐述写信的原因:
  I am writing to …
  I am writing concerning/regarding/with regard to/with reference to/about….

- Referring to a previous contact 提及之前的联系:
  Further to your letter of /dated…
  I am writing to thank you for your letter of/dated…

- Asking for information 索取信息:
  I would be grateful if you could send me…
  I would appreciate it if…

- Giving information 提供信息:
  Here is the information you wanted.
  We are pleased to inform you that…

- Confirming information 确认信息:
  I am happy to write to confirm our agreement about…
  I am able to confirm that…

- Requesting 有礼貌地询问:
  I would be grateful if you could…
  We would be delighted if you could…

- Booking 预订:
  I would like a room with a bathroom, from…to…
  Would it be possible for me to have a room at the back of the hotel?

- Ordering 订货:
  I would like to order…
  We would like to place the following order:

- Reminding 提醒:
  May I remind you that…?
  We would like to remind you that your remittance was due on 30th August, 2008.

- Complaining 抱怨:
  We are disappointed to find that the quality of the goods you supplied was poor.

I am writing to complain about the punctuality, quality and customer service on…

- Warning 警告:

I must ask you either to send my order immediately or to refund my payment of ￥7000.

If you do not…, we will cancel our order.

- Apologizing 道歉:

I would like to apologise for…

Please accept my apology for…

I am very sorry that I was unable to…

- Promising 承诺:

I can assure you that we will...

I promise to give the matter my immediate attention.

- Suggesting 建议:

We propose…

It is proposed/suggested/advised that…

- Offering 提议/给予:

We offer you the easiest repayment terms.

If you require any further information/details, please do not hesitate to contact us.

- Inviting 邀请:

We should like to invite you…

We should be delighted if you could…

- Accepting 接受（邀请）:

Thank you very much for your kind invitation to…I would be very pleased to go.

It was very kind of you to invite me to…I should be delighted to go.

- Declining 拒绝（邀请）:

It was very kind of you to invite me but I am afraid I can not attend because…

Unfortunately, due to…, I am unable to…

- Thanking 感谢:

Thank you for your contribution to the project.

I would like to express my appreciation for your…

- Hoping 希望:

I hope I shall not be obliged to take this matter any further.

I do hope your schedule will allow you to accept our invitation.

- Referring to future contact 提及将来的联系或合作:

I look forward to hearing from you.

I look forward to seeing you.

- Complimentary close 礼貌的结尾:

Sincerely yours (typical, less formal) = Yours sincerely

Very truly yours (polite, neutral) = Yours truly

Sample Letter 1:

Mrs. Zhou Qian
Sales Manager
Fine Computers
68 Nazhi Rd
Harbin
June 18th, 2020

Mr. Liu Mei
59 Huangjiang St
Harbin

Dear Mr. Liu

I am delighted that you are interested in our new lap-top computer advertised on Harbin Daily yesterday.

Attached is the catalogue for our new range of products with a price list. Please contact me if you have any questions.

I look forward to receiving your order.

Yours sincerely
Zhou Qian

Sample Letter 2:

Mr. Wang Min
Personnel Manager
Viewtech Company
35 Tiandi St
Harbin
Sep 15th, 2020

Mr. Gao Junguang

59 Liaoyuan St
Zhengzhou

Dear Mr. Gao

Thank you for applying for the position offered by our company.

We arranged an job interview for you on September 18th, 2020. Please arrive at 9:30 in the morning. You also need to take your CV and two reference letters with you.

The best way to come to our office is to take the direct Bus No. 219 at the Bridge Bus Station and get off at Tiandi Bus Stop.

If you have any questions, please contact us.

Yours sincerely
Wang Min

Exercise 7

Fill in the gaps with *Writer's address, Receiver's address, Date, Salutation, Body, Complimentary close, Signature.*

1 _____
2 _____

3 _____

4 _____

5 _____

6 _____

7 _____

Transcript

Listening 1 Transfer calling

(ring~~~~)

A: Good morning. Creative Software Manager's Office. How can I help you?

B: Oh, good morning. This is Wang Qiang from Advanced Micro Devices. May I speak to Ms. Liu, please?

A: I am sorry but she's not in right now. If you give me your name and number, I'll ask her to call back.

B: OK. My name is Wang Qiang. My number is 010-6654×××related×.

A: I see. I'll tell Ms. Liu you called as soon as she's in.

B: Thank you. Goodbye.

A: You are welcome. Thank you for calling.

Listening 2 Leaving message

(ring~~~~)

A: Good afternoon. Creative Software Corporation. How can I help you?

B: Good afternoon. This is Emily speaking. May I speak to Kate Johnson, please?

A: I'm sorry, but she is on another line. Could you hold on?

B: Well, it's not very urgent. Could you please take a message?

A: No problem. Please give me your name and number.

B: Thank you. My name is Emily. My number is 010-55241686. I just wanted to discuss activity arrangements with her.

A: I'll give her your message as soon as she's available.

Unit 4 Creative Software

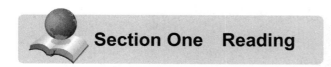

Section One Reading

➢ Technical Reading

Using a personal computer or workstation to produce high-quality printed documents. Desktop publishing systems have become increasingly popular for producing newsletters, brochures, books, and other documents that formerly required a typesetter.

Desktop Publishing

"Desktop publishing" refers to the use of personal computers to design, implement and publish books, newsletters, magazines and other printed pieces. Desktop publishing is really a combination of a few different processes including word processors, graphic design, information design, output and repress technologies, and sometimes image manipulation. There are also many applications that support these processes, including font creation applications (that allow users to design and create their own typefaces, called fonts) and type manipulation applications (that allow users to modify text in visually creative ways).

A desktop publishing system allows you to use different typefaces, specify various **margins** and **justifications**, and embed **illustrations** and graphs directly into the text. The most powerful desktop publishing systems enable you to create illustrations, while less powerful systems let you insert illustrations created by other programs.

As word-processing programs become more and more powerful, the line separating such programs from desktop publishing systems is becoming blurred. In general, though, desktop publishing applications give you more control over typographical characteristics, such as kerning, and provide more support for full-color output.

A particularly important feature of desktop publishing systems is that they enable you to see on the display screen exactly how the document will appear when printed. Systems that support this feature are called WYSIWYGs (what you see is what you get).

Desktop publishing systems have become increasingly popular for producing newsletters, brochures, books, and other documents that formerly required a typesetter. Once you have produced a document with a desktop publishing system, you can output it directly to a printer or you can produce a **PostScript** file which you can then take to a service bureau. The service bureau has special machines that convert the PostScript file to film, which can then be used to make plates for offset printing. Offset printing produces higher-quality documents, especially if color is used, but is generally more expensive than laser printing.

Vocabulary

1. margin   n. 页边距
2. justification   n. 对齐方式
3. illustration   n. 图解、插图
4. PostScript   n. 页面描述语言

Exercise 1

Read the article and find answers to these questions.

1. A particularly important feature of desktop publishing systems is called _____.
2. Service bureau offer services such as _____.
3. PostScript fonts were created by _____.
4. Fonts refers to the style and size of a _____.

➢ Fast Reading

Computer Graphics

Computer graphics are graphics created using computers and, more generally, the representation and manipulation of image data by a computer with help from specialized software and hardware.

The development of computer graphics has made computers easier to interact with, and better for understanding and interpreting many types of data. Developments in computer graphics have had a profound impact on many types of media and have revolutionized animation, movies and the video game industry.

Creating computer graphics requires a digital computer to store and manipulate images, a display screen, input/output devices, and specialized software that enables the computer to draw, color, and manipulate images held in memory. Common computer graphic formats include GIF and JPEG, for single images, and MPEG and Quicktime, for multiframe images. The field has widespread use in business, scientific research, and entertainment. Monitors attached to CAD/CAM systems have replaced drafting boards. Computer simulation using graphically displayed quantities permits scientific study and testing of such phenomena as nuclear and chemical reactions, gravitational interactions, and physiological systems. computer animation; computer art.

In the process the computer uses hundreds of mathematical formulas to convert the bits of data into precise shapes and colors. Graphics can be developed for a variety of uses including presentations, desktop publishing, illustrations, architectural designs and detailed engineering drawings.

Mechanical engineers use sophisticated programs for applications in computer-aided design and computer-aided manufacturing. Let us take, for example, the car industry. CAD software is used to develop, model and test car designs before the actual parts are made, this can save a lot of time and money.

Today, three-dimensional graphics, along with colour and animation, are essential for such applications as fine art, graphic design, Web-page design, computer-aided engineering and academic research. Computer animation is the process of creating objects and pictures which move across the screen; it is used by scientists and engineers to analyze problems. With the appropriate software they can study the structure of objects and how it is affected by particular changes.

Basically, computer graphics help users to understand complex information quickly by presenting it in a clear visual form.

Exercise 2

Read the article and find answers to these questions.
1. What are "computer graphics"?
2. What do the acronyms "CAD", "CAE" and "CAM" stand for?
3. What are the benefits of using computer graphic in the car industry?
4. What are the benefits of using graphics in business?

➢ Supplementary Reading

Multimedia Magic!

Until now multimedia applications have been used mainly in the fields of information, training and entertainment. For example, some museums, banks and estate agents have information kiosks that use multimedia. Several companies produce training programmers on optical disks, and marketing managers use presentation packages (like Microsoft PowerPoint or Lotus Freelance Graphics for Windows) to make business presentations. They have all found that moving images, sound and music involve viewers emotionally as well as inform them, and make their message more memorable.

The word multimedia is a combination derived from multiple and media. The computer is an intrinsic part of multimedia. Thus, all the elements of multimedia have to be in digital format. This chapter provides an overview of multimedia systems. The multimedia computer has the capability to play sounds, accurately reproduce pictures, and play videos—now easily available and widely in use. The development of powerful multimedia computers and the evolution of the Internet have led to an explosion of applications of multimedia worldwide. Multimedia systems are used for education, in presentations, as information kiosks, and in the gaming industry. To use it effectively, one has to understand not only how to create specific elements of multimedia, but also to design the multimedia system so that the messages are conveyed effectively. It is important to be sensitive to other multimedia—such as TV and films—to create effective multimedia. Several elements of multimedia system consists of sound, graphics, text, and video. The multimedia production and the stages involved in it are also described.

A multimedia system is characterized by computer-controlled, integrated production, manipulation, presentation, storage and communication of independent information, which is encoded at least through a continuous (time-dependent) and a discrete (time-independent) medium.

Multimedia involves multiple modalities of text, audio, images, drawings, animation, and video. Examples of how these modalities are put to use:

1. Video teleconferencing.

2. Distributed lectures for higher education.

3. Tele-medicine.

4. Co-operative work environments.

5. Searching in (very) large video and image databases for target visual objects.

6. "Augmented" reality: placing real-appearing computer graphics and video objects into scenes.

7. Including audio cues for where video-conference participants are located.

8. Building searchable features into new video, and enabling very high- to very low-bit-rate use of new, scalable multimedia products.

9. Making multimedia components editable.

10. Building "inverse-Hollywood" applications that can recreate the process by which a video was made.

11. Using voice-recognition to build an interactive environment, say a kitchen-wall web browser.

Exercise 3

Match these terms in the box with the explanations:

| 1. Computer animation | 2. Video computing | 3. MIDI interface |
| 4. CD-ROM player | 5. Multimedia control panels | |

a. Manipulating and showing moving images recorded with a video camera or captured from a TV or video recorder.

b. Images which move on the screen.

c. Small programs inside the PS designed to work with audio and video files.

d. A drive used to handle CD-ROM disks.

e. A code for the exchange of information between PCs and musical instruments.

Section Two   Listening

Section Three   Speaking

Exercise 4

What would you say in the following situations?

1. Imagine that you work at Macromedia company. You are answering a phone call. Play your role according to the clues given in the brackets. （假设你在多媒体软件公司工作。现在你正在接听电话。根据下面括号中给出的提示完成角色扮演对话。）

You:（电话铃响起，您接听电话，并说出 Macromedia 的名称。）

Client:（自我介绍。说明自己正在使用的 Dreamweaver 的版本，并询问如何设置页面的背景图片。）

You:（告诉对方相关操作的大致步骤。）

Client: Oh, I see. Thanks a lot!

You:（表示客气，并祝对方愉快。）

2. Imagine that you work at Discreet company. You are answering a phone call. Play your role according to the clues given in the brackets. （假设你在 Discreet 公司工作。现在你正在接听电话。根据下面括号中给出的提示完成角色扮演对话。）

You: （电话铃响起，您接听电话，并说出 Discreet 的技术支持部。）

Client:（自我介绍。说明自己正在使用 3D Studio Max7，每次启动大约 5 分钟后就会死机，询问如何解决。）

You: （询问对方电脑使用的显卡型号，并提示将显卡设置为 32 位。）

Client: Let me see. Yes, it is. Is this the cause of the problem?

You: We will see. Now set the display card to 32 bit. This may solve the problem.

Client: Thanks. It is done.

You:（表示客气，并祝对方顺利。）

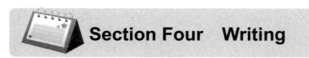

Writing a report 报告的写作

A report is a very formal document that is written for a variety of purposes. It should be noted that reports are considered to be legal documents in the workplace and, thus, they need to be precise, accurate and difficult to misinterpret.

报告是一种非常正式的文件，有各种不同的种类（由不同目的决定）。值得注意的是，在工作场合，报告被看成是具有法律效力的文件，因此报告的写作应该简洁、准确、没有歧义。

A report should include the following points: 报告应该包括以下内容：

1. Title 标题（关于什么的报告）
2. Introduction 介绍（报告写作的目的）
3. Findings （调查）发现（发现的问题和相关背景数据）
4. Conclusion 归纳总结（分析总结出关键问题）
5. Recommendations 建议（提出解决问题的建议或办法）

Layout of a report 报告写作格式

Report on …

Introduction   (introduce the aim of the report)

This report aims/sets out to…

The aim/purpose of this report is to…

This report is based on…

Findings (what problem or other information is found)

It was found that…

The following points summarise our key findings.

The key findings are outlined below.

Conclusion (draw a conclusion)

It was decided/agreed/felt that…

It is clear that…

No conclusions were reached regarding…

Recommendations (give suggestions)

It is suggested/proposed/recommended that…

There are a number of reasons for…

There are several factors which affect…

A further factor is…

This raises a number of issues.

As might have been expected, …

Contrary to expectations, …

Sample 1:

Report on Working Conditions

Introduction

The aim of this report is to assess the main reasons for staff complaints about working conditions and propose ways of improving the situation. It is based on the results of a detailed questionnaire sent to all employees in addition to interviews with managers and staff representatives.

Findings

As might have been expected, low salary is the key reason for staff complaints. Moreover, a large number of employees are not satisfied with the current level of bonus and benefits. Another problem is the employees' working environment. In particular, poor ventilation and lighting in communal areas such as the canteen and coffee room have been highlighted.

Conclusion

It is clear that pay, ventilation, lighting and communal areas are the specific problems which need to be solved.

Recommendations

In order to deal with the issue of pay, it is recommended that a meeting should be held to discuss pay levels. In addition, bonus, rewards, or fringe benefits such as subsidized health care should be entitled to employees having been with the company for over two years. It is also suggested that a Suggestions Box should be put in the canteen to collect solutions to working environment.

Sample 2:

Cost-cutting: Administration Department

Introduction

The aim of this report is to examine ways of cutting costs in the Administration Department and explain the implications of these cuts for the running of the department. It is based on the results of a detailed questionnaire sent to all employees.

Findings

It is clear that within the department there are a number of areas where cost-cutting measures could be taken. The most important areas of concern are the following:

- paper
- refreshments

Recommendations

In order to solve the problem of paper, it is suggested that the department installs a system to recycle all used printing and photocopying paper. It is expected that by adopting new recycling procedures, the department could save as much as £120 a month.

As for refreshments, it is recommended that tea and coffee should only be supplied free to employees during morning and afternoon breaks. The measure should help save £150 a month.

Conclusion

It is felt that the above measures will result in a lot of savings for the Administration Department. Although these recommendations are not expected to affect the running of the department in any way, managers should be prepared to encounter initial resistance from staff.

Sample 3:

Worldnet Report

Introduction

The purpose of this report is to evaluate Internet services offered by Worldnet, a chain of 24-hour Internet café in London. There has been a steady decrease in the number of customers over the past six months. This report will analyse the trend based on findings from a survey. It will assess the present services and conclude recommendations for improvements.

Findings

Worldnet satisfies a wide range of customers with different backgrounds and ages. However a large number of customers commented that the cost of the service was too high. The speed of the Internet connection was said to be extremely slow and unstable at peak times.

In terms of the hardware, the computer terminals were generally thought to be out of date and badly maintained. On a more positive note, the café was considered to be good value. Customers are satisfied with the quality of the food but there were some complaints about the lack of variety.

Recommendations

In order to become more competitive, Worldnet should introduce new half-hourly rates immediately. Special student rates should also be considered as this would attract 16-25-year-old customers, who represent the majority of Internet users. Furthermore, a faster and more reliable service is highly recommended. The computer terminals and chairs also need upgrading and maintaining more regularly. The café should also provide a wider range of food and drinks. If these recommendations are put into practice, the number of Worldnet customers should start to increase substantially.

Exercise 5

Write a report assessing the suitability of your job (Resource Planning Manager) for home-based working.

Transcript

Listening 1

Kevin: Hello, this is Adobe Systems Customer Service. What can I do for you?

Customer: Hello, something is wrong when I start the Photoshop CS2 program I just bought. It prompts "msvcrt11.dll could not be found and reinstalling might solve the problem."

Kevin: Oh, you are probably using plug-in for older versions of Photoshop CS2.

Customer: Then what can I do to solve this problem?

Kevin: You can solve it by removing all the original and third-party plug-ins designed for older versions of Photoshop CS from the Plug-ins folder.

Customer: Sorry, I don't think I have this problem. I bought your latest product, and this is my first time to install it. Any other suggestions?

Kevin: In that case, you must have selected the customized mode, haven't you?

Customer: Oh, yes, I have.

Kevin: If this is your first time using this software, we suggest you use the full installation mode so that all the dlls can be installed to the Windows system folder and then you will not have any missing components.

Customer: OK, I got it. Thanks for the help.

Kevin: You're welcome. Thank you for using Adobe Systems products and services. Goodbye.

Listening 2

Kevin: Hello, Autodesk Support Center. Can I help you?

Customer: Hello, this is Jimmy speaking. I am using AutoCAD 2007 to handle a sophisticated project blueprint. I want to delete an empty layer, but it won't work. What's the problem?

Kevin: You should be able to delete an empty layer. Is there any other layer that has references to it?

Customer: I don't think so.

Kevin: Well, if you are sure that all the objects and references from other layers are removed, this layer is probably frozen in a AutoCAD view port. Unfreeze it, and you can delete it.

Customer: Ah, yes. I got it. That will solve the problem. Thank you.

Kevin: You're welcome. Thank you for using our products and services.

Unit 5   Communicate Online

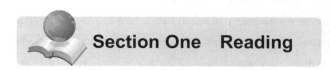 Section One   Reading

➢ Technical Reading

These are three different ways of communicating with other people over the Internet.

Online Communication

E-mail (short for electronic mail) is software you use to electronically communicate with other people. You send an E-mail message instead of writing a letter and sending it through the postal system. Using E-mail is almost as simple as using a **Web browser**. To receive E-mail messages, you need an E-mail address. An E-mail address is a **unique** address for a person using an E-mail system. Consider Stephen Green, a student at UCLA. His E-mail address could appear as Stephen_green@ucla.edu. Not all E-mail systems use a person's full first and last name; some will simply use the first initial with the last name such as sgreen@ucla.edu.

One way to send or receive E-mail is to use a mail program such as Outlook Explorer that will communicate with your mail server. Another method is called Webmail (such as Hotmail or Gmail) that allows you to send messages using a Web-based interface.

As the Internet became more popular in homes and offices, online communication also began to take off. Initially people were fascinated by electronic mail, which allowed users to send a virtual message to anyone, anywhere. This eliminated the need for handwritten correspondence, as well as **drastically** decreasing the amount of time it took to send and receive communication. It also was a free service, saving people thousands on postage, paper, and other supplies necessary to produce hard copies of correspondence.

Now, many people receive E-mails on their mobile phones, giving them access anywhere they may be located. Chat rooms and instant messaging was also a popular advancement in online communication, allowing friends and family members to communicate instantly and securely with just an internet connection. Individuals using chat rooms could meet people from all over the world with a shared interest in a specific topic, and communicate privately without having to reveal any personal information.

Social networking has taken the world by storm in the past years, with the creation of sites such as **Facebook**, **Myspace**, **Twitter**, and more. These sites allowed users to create and maintain personal profiles, and keep in contact with hundreds of friends and family members in one place.

Initially Facebook was created for college students, but now has billions of users worldwide of all ages. Twitter has allowed people to see the thoughts and opinions of celebrities, stay in touch with current events, and share their own ideas to anyone who follows. Most businesses now advertise on social media sites, knowing this is where they can reach the largest number of people. The creation of online communication has made it so much easier to stay in touch with those who are far away, as well as view photos and videos, and stay on top of birthdays, anniversaries, and other meaningful events in the lives of others.

Vocabulary

1. Web browser   *n.* 网络浏览器
2. unique   *adj.* 唯一的；独一无二的
3. drastically   *adv.* 大大地，彻底地；激烈地
4. Facebook   *n.* 美国的一个社交网络服务网站
5. Myspace   *n.* 聚友网
6. Twitter   *n.* 国外的一个社交网络及微博客服务网站

Exercise 1

True or False

1. (   ) E-mail is the exchange of computer-stored messages by telecommunication.
2. (   ) The only way to send or receive E-mail is to use a mail program that will communicate with your mail server.
3. (   ) People can receive E-mails on their mobile phones.
4. (   ) An E-mail address must have letters and dots and an "@" (meaning "at").

➢ Fast Reading

iPad-User's Guide－Mail

Write messages

Mail lets you access your E-mail accounts, on the go.

Unit 5 Communicate Online

Change the preview length in
Settings→Mail, Contacts, Calendars

Insert a photo or video. Tap the insertion point, then tap Insert Photo or Video.

Quote some text when you reply. Tap the insertion point, then select the text you want to include. Tap, then tap **Reply**. You can turn off the indentation of the quoted text in Settings→Mail, Contacts, Calendars→Increase Quote Level.

Send a message from a different account. Tap the From field to choose an account.

Change a recipient from Cc to Bcc. After you enter recipients, you can drag them from one field to another or change their order.

Get a sneak peek

See a longer preview. Go to Settings→Mail, Contacts, Calendars→Preview. You can show up to five lines.

57

Is this message for me? Turn on Settings→Mail, Contacts, Calendars→Show To/Cc Label. If the label says Cc instead of To, you were just copied. You can also use the To/Cc mailbox, which gathers all mail addressed to you. To show it, tap Edit while viewing the Mailboxes list. Finish a message later.

Save it, don't send it. If you're writing a message and want to finish it later, tap Cancel, then tap Save Draft.

Finish a saved draft. Touch and hold Compose. Pick the draft from the list, then finish it up and send, or save it again as a draft.

Show draft messages from all of your accounts. While viewing the Mailboxes list, tap Edit, tap Add Mailbox, then turn on the All Drafts mailbox.

Delete a draft. In the Previous Drafts list, swipe left across a draft, then tap Delete.

Gather important messages. Add important people to your VIP list, and their messages all appear in the VIP mailbox. Tap the sender's name in a message, then tap Add to VIP. To show the VIP mailbox, tap Edit while viewing the Mailboxes list.

Get notified of important messages. Notification Center lets you know when you receive messages in favorite mailboxes or messages from your VIPs. Go to Settings→Notification Center →Mail.

Flag a message so you can find it later. Tap while reading the message. You can change the appearance of the flagged message indicator in Settings→Mail, Contacts, Calendars→Flag Style. To see the Flagged smart mailbox, tap Edit while viewing the Mailboxes list, then tap Flagged.

Search for a message. Scroll to or tap the top of the message list to reveal the search field. Searching looks at the address fields, the subject, and the message body. To search multiple accounts at once, search from a smart mailbox, such as All Sent.

Search by timeframe. Scroll to or tap the top of the message list to reveal the search field, then type something like "February meeting" to find all messages from February with word "meeting".

Search by message state. To find all flagged, unread messages from people in your VIP list, type "flag unread vip". You can also search for other message attributes, such as "attachment".

Junk be gone! Tap while you're reading a message, then tap Move to Junk to file it in the Junk folder. If you accidentally move a message, shake iPad immediately to undo.

Make a favorite mailbox. Favorite mailboxes appear at the top of the Mailboxes list. To add a favorite, view the Mailboxes list and tap Edit. Tap Add Mailbox, then select the mailbox to add. You'll also get push notifications for your favorite mailboxes.

Attachments

Save a photo or video to your Camera Roll. Touch and hold the photo or video until a menu appears, then tap Save Image.

Use an attachment with another app. Touch and hold the attachment until a menu appears, then tap the app you want to open the attachment with.

See messages with attachments. The Attachments mailbox shows messages with attachments from all accounts. To add it, view the Mailboxes list and tap Edit.

Exercise 2

Read the information and answer the following questions.
1. Can users use one more account to send and get mail?
2. What can you do to save the draft when you're writing a message and want to finish it later?
3. How to gather important messages?
4. How to make a favorite mailbox?

➢ Supplementary Reading

Text 1

E-mail

You've seen how simple it is to send an E-mail message, even simpler than sending a letter and going through postal mail. E-mail is fast, immediate and free. Perhaps you've noticed – while composing a message – Bcc and Cc buttons and wondered about them. To send an E-mail message you need to type an E-mail address next to.

Cc and Bcc have similar functions, even more interesting.

Cc (Carbon Copy) allows you to send copies of the same message to multiple E-mail addresses.

Bcc (Blind Carbon Copy) allows you to send messages to people without including others E-mail addresses in the message header.

With To and Cc all E-mail addresses appear in the header of the message you send. If you frequently send messages to lots of people, use Bcc instead of Cc and no one will be able to know to whom you're sending the message.

Tip: you can easily add an E-mail by clicking on (besides To, Cc, Bcc).

If you are composing a long E-mail message, click File→Save, and a copy of your message is immediately saved in Draft folder. That way you will never lose your messages if something bad happens. Press on Draft folder, open the message and continue from where you left. You can also save messages to any location on your computer. Select a message from a folder (e.g. Inbox), double-click to open it and click File→Save As… specify a location and press Save.

When saving you can choose between three different formats:

Mail format (.eml): The file is saved as. eml, 2-click its icon to launch Outlook Express.

TXT Format (.txt): message is stored as a text file, read with Notepad or any text editor.

Html format (.htm,. html): This format is not available for all messages, it allows to save a message as webpage.

The Mail format requires Outlook Express to open the message.

TXT is the safest, since the message can be viewed using any text editor e.g. NotePad.

Html messages can be viewed within any browser e.g. Internet Explorer or Netscape Navigator. If you want to exchange messages through floppies use TXT format.

In most of the time you will be using the Address Book to add E-mail addresses as recipients. In case you've typed an E-mail address manually, click to verify that the address is correct or check if it's available in address book, if not simply click on New Contact in the Check Names window to add the new entry to your Address book.

Exercise 3

Read the information and answer the following questions.

1. Bcc differs from To and Cc in that _____.

   A. it's faster and simpler

   B. it can send message to multiple receivers

   C. it won't reveal others E-mail address

2. If you want to save your message as a webpage, you should choose the format of _____.

A. .eml
B. .txt
C. .htm

3. If you have filled in an address manually, you may click to _____.

A. check if the address is correct
B. add a new entry to your Address book
C. change or delete the address

Text 2

How do I publish photo descriptions on Facebook using iPhoto?

Apple iPhoto 11 (version 9) includes a "Share" feature that allows you to publish photos directly to Facebook. You can either create new albums or add photos to existing albums. The upload process is pretty straightforward, but adding descriptions (captions) to each photo is not.

When you publish photos to Facebook using iPhoto, you may notice that all your images have descriptions like IMG_1760, IMG_1761, etc. This is because when you import images into iPhoto, the program automatically uses the filename (minus the file extension) as the image title. When iPhoto publishes your photos to Facebook, it uses the Title field for the description of each published image.

This can be confusing, since iPhoto also includes a Description field. You can access both the Title and Description fields by selecting View → Info while viewing a photo. If you add a description where it says "Add a description…", your Facebook photo caption will include the title, followed by two line breaks, then the description.

Therefore, the best way to caption your Facebook photos is to simply replace the title with the caption in iPhoto before you publish them. You can leave the description field blank unless you want to publish a long description for the photo. Following this tip will save you the time of manually updating all your captions after you publish your photos.

Section Two Listening

Listening 1

Referring to a User Manual

这星期，Kevin 接受了有关如何委婉地回答问题的培训。Lily 告诉他："如果问题过于笼统，你最好给询问者提供一个参考，比如说请他查询在线常见问题列表。"Kevin 点点头。这时电话铃响了……

Exercise 4

Listen to the dialogue again and fill in the notes outline.

1. Chen Lin is calling _____ about _____.
2. The network adapter in the computer of Chen Lin is _____.
3. Kevin considered the question too _____ and not directly answer.
4. Finally, Kevin _____ to solve the problem.

Listening 2

Creating a Wireless Connection

Kevin 是一家电脑公司客户服务部门的员工，一天他接到一位客户的电话，咨询如何创建无线网络连接。他该如何答复呢？

Dictation

Products' models and Key Terms

(Ring~~~~~)

A: _____. Can I help you?

B: Good morning. I _____ that you are recalling your _____.

A: Yes.

B: I'd like to know if the battery of my computer is _____.

A: What model are you using?

B: Latitude _____.

A: Will you tell me the _____?

B: _____? What is a service tag?

A: It is a kind of serial number. It is on the back of your computer.

B: Yes, I have found it. It is _____.

A: Could you tell me your battery serial number?

B: It is _____.

A: I'm sorry. Your battery is not on the recall list.

B: Sorry to hear that. But my battery does not last very long.

A: Well, a _____ can be affected by many factors, such as some drives and software which are power consuming.

Unit 5　Communicate Online

B: _____. Thank you very much.
A: You are welcome.

 Section Three　Speaking

Useful expressions: finding out problems

Exercise 5

Jason has received a call form a client, Mr. Jones wants to know how to active his Windows system. What would he say in this situation?（Jason 接到客户 Jones 先生的电话，他希望知道如何激活他的 Windows 系统。在这种情形下他将怎么说？）

Jason：_____. Creative Software. How _____ ?
Mr. Jones: Oh, good morning. This is Corrin Jones. I want to ask you how _____ ?
Jason：Well, _____, and then _____.
Mr. Jones: That's all? _____.
Jason：_____.

 Section Four　Writing

E-mail 电子邮件

E-mail is an electronic mail.电子邮件是电子信件。An **E-mail letter** is a letter which is sent as an E-mail using a computer. 一封电子邮件是用电脑发送的电子信件。It is a new means of communication between friends or organizations.电子邮件是朋友之间或机构之间一种新的相互交流方式。

An E-mail can be formal or informal depending on whom you write to. 电子邮件的文体正式与否取决于收件人是谁。To someone you don't know, make it a bit formal; 写给陌生人，文体要正式一些。To someone you know well, make it less formal or informal.写给熟悉的人，文体可以不用特别正式。

Important Points to Remember 谨记下列事项
1. E-mail is less formal than a written letter. 电子邮件没有书面信函的文体正式。

2. E-mails are usually short and concise. 电子邮件通常言简意赅。
3. When writing to someone you know well, feel free to write as if you are speaking to the person. 当给熟人发电子邮件时，邮件读起来应该就像是和收件人说话一样。
4. Use abbreviated verb forms (He's, We're, He'd, etc.)
   可以使用缩写，比如 He's、We're、He'd 等。

Layout of an E-mail 电子邮件格式

<u>Header fields 电子邮件信头</u>

The message header usually includes at least the following fields:

From: langskill@hotmail.com.cn (the E-mail address, and optionally the name of the sender)

To: fancyworld@sohu.com.cn (the E-mail addresses, and optionally names of the message's recipients)

Subject: welcome to visit my company (a brief summary of the contents of the message)

Date: April 20th, 2020 (the local time and date when the message was written)

CC: send a copy to…

BCC: send a blind copy to…(the other people don't know you are sending this copy)

<u>Content 内容</u>

1. **Greeting:** Dear Mr…, (formal); Hello, (informal)
2. **Body:**
   Opening paragraph
   Main body
3. **Complimentary close:** Yours sincerely, (formal); Yours, (informal)
4. **Signature:** name and position

For Example:

E-mail to inquire about people's preferences

From: yukichen@126.com
To: customersun@gmail.com, jerry@yeah.net, tony@hotmail.com, jim@yahoo.com
CC:
BCC:
Subject: inquiry about your preferable food for the banquet
Date:
Dear All, As you are invited to our company banquet on Sep 10th, 2020. We would like to know your preferences.

Please let me know the following information:

Do you prefer Western food or Chinese food?

Are you a vegetarian?

Are you fond of spicy, hot food?

Is there anything that you would like to order for the meal?

Please don't hesitate to contact us if you have special requirements. I look forward to your early reply.

Yours,
Yuki Chen
Software Engineer

From: mariasanderson@freemail.com
To: wenruihuang@gmail.com
CC:
BCC:
Subject: build our research project team
Date:

Dear Jonny,

You are arranged to be a member of our software development project team because you are good at Java.

Please get ready to join the team. You will be responsible for using your Java knowledge and your previous work experience to help check the progress of our research project.

I would like to know how you think of it.

I look forward to your reply.

Yours,
Maria Sanderson
Software Team Head

From: jellygao@yahoo.com
To: helenzhang@hotmail.com
CC:
BCC:
Subject: introduction of 2 team members
Date:

Dear Helen,

I am writing to introduce to you two project team members we are sending to you next week.

The first one is Wang Jianhong, male, who is good at ASP. Net. He has been working in project teams on software development for 4 years.

The second one is Zhao Jie, female, who is excellent and experienced at Java. She has 5 years experience in using Java to programme.

I hope these two members will meet your need.

Please keep in touch.

Yours,
Jelly Gao
Personnel Assistant

Exercise 6

You work for General Computers Co. Ltd. You are supposed to write an E-mail to your customer asking for their preferences for the color of the notebooks he ordered.

From:
To:
CC:
BCC:
Subject:

Unit 5　Communicate Online

Date:

Transcript

Listening 1　Referring to a User Manual

(Ring~~~~~)

Kevin:　　　Creative Software. Hello, what can I do for you?

Customer:　Hello, this is Chen Lin. I just bought a personal computer produced by your company. But I have no idea how to access the Internet. Can you help me?

Kevin:　　　Do you have any problem with the network adapter or something?

Customer:　No, there is noting wrong. By the way, what is a network adapter? I have no idea how to use my computer to access the Internet. Can you tell me how to do that?

Kevin:　　　If you want to know about the fundamentals of network operations, I would advise you to read our Users' Manual first. There are detailed explanations in it.

Customer:　Well, if you say so, I will read that first. Thanks.

Kevin:　　　You're welcome. Thank you for you calling.

Listening 2　Creating a wireless connection

(Ring~~~~~)

Kevin:　　　Hi, Creative Software Corporation Customer Support. Can I help you?

Customer:　Hi, Good morning. I am trying to use the wireless LAN on my Windows XP system. But I can't access the network. Please tell me how to enable the connection.

Kevin:　　　You should first check whether your operating system has recognized your WLAN adapter. Check it in your Device Manager.

Customer: Oh, I checked. Everything is fine.

Kevin: Then you should try to configure it the correct way. Open Properties interface for TCP/IP in your adapter Properties page, and let the system automatically obtain its IP and DNS address.

Customer: I understand. I will try to do that. Thanks.

Kevin: Anytime. Thank you for your calling.

Dictation

(Ring~~~~~)

A: <u>Dell Technical Support</u>. Can I help you?

B: Good morning. I <u>read from the newspaper</u> that you are recalling your <u>computer batteries</u>.

A: Yes.

B: I'd like to know if the battery of my computer is <u>on the recall list</u>.

A: What model are you using?

B: Latitude <u>D610</u>.

A: Will you tell me the <u>service tag</u>?

B: <u>Beg your pardon</u>? What is a service tag?

A: It is a kind of serial number. It is <u>on the back of your computer</u>.

B: Yes, I have found it. It is <u>3TYJP1X</u>.

A: Could you tell me your battery serial number?

B: It is <u>JP-0Y1338-48630-680-0749</u>.

A: I'm sorry. Your battery is not on the recall list.

B: Sorry to hear that. But my battery does not last very long.

A: Well, a <u>battery's life</u> can be affected by many factors, such as some drives and software which are power consuming.

B: <u>That's very kind of you</u>. Thank you very much.

A: You are welcome.

Unit 6   Surf the Network

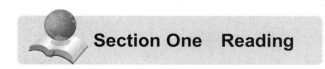

Section One   Reading

➢ Technical Reading

The Web is exploding all around you. Never before has the world seen such a dynamic and exciting technology become a part of our lives so quickly. Many young people growing up today cannot even comprehend life without the Web. Reading the following articles can help you to understand the network better.

World Wide Web Overview

Simply, the World Wide Web is a way to share resources with many people at the same time, even if some of those **resources** are located at opposite ends of the world.

- **TCP/IP**

TCP/IP (Transport Control Protocol/Internet Protocol) is the basic communications protocol that makes the Internet work. It defines the rules that allow various computers to communicate across the Internet. It doesn't matter if you're viewing a multimedia presentation from a Web server, transferring a file from an FTP server, or chatting across an IRC server, TCP/IP is the foundation for the movement of the information.

- **Hypertext Transfer Protocol (http)**

Hypertext transfer protocol (http) is the communications protocol that supports the movement of information over the Web, essentially from a Web server to you. That's why Web site addresses start with "http://". That beginning portion of the address informs all the Internet technologies that you want to access something on the Web. For example, it tells your browser software that you want to access a Web site. Most browser software today assumes that you want to access a Web site on the Internet. So, you don't even have to type in the "http://" if you don't want to.

If you recall, the Web is the Internet in multimedia form. So, when you access a Web site, your computer is using both TCP/IP (to transfer any type of information over the Internet) and http (because the information you want is Web-based).

- **File Transfer Protocol (FTP)**

File transfer protocol (FTP) is the communications protocol that allows you to transfer files of information from one computer to another. FTP servers are popular stops for Web surfers who want to download a variety of different types of files—software such as games and music are the most common. When you download a file from an FTP site, you're using both TCP/IP (the basic Internet protocol) and FTP (the protocol that allows you to download the file).

Web pages can include text information, pictures, sounds, video, FTP links for **downloading software**, and much more. You can create living **documents** that are updated weekly, daily, or even hourly to give web surfers a different experience every time they visit your pages. As the technology develops, even more amazing applications will be possible.

- **Understanding URLs**

The URL, or Uniform Resource Locator, is the address on the web that you are visiting. If someone gives you the address to their web page, they may say its at "www.imagescape.com". Most web browsers need you to include **http**:// at the beginning of the URL so the program knows that you want to visit a web page. Remember, you can also connect to FTP sites and gophers with your browser, so you need to be specific FTP sites.

- **Web page**

Each day when browsing the Internet, we visit a lot of websites, some more complex, others - just simple personal pages. The term "website" represents a summary of all the content you have put online - each file takes part in what the website represents. And the driving power behind the website, the pillars that hold it together, are the web pages.

Each web page (also known as webpage) represents various types of information presented to the visitor in an aesthetic and readable manner. Most of the web pages are available on the World Wide Web, which makes them widely accessible to the Internet public. Others may be also available online but only restricted to a certain private network, such as a corporate intranet. The information in all those web pages is located on remote web servers in the form of text, image, or script files. A smaller amount of web pages are intended for home or test use and are located on local computers, which do not need Internet connection to display them.

The information on a web page is displayed online with the help of a web browser, which connects with the server where the website's contents are hosted through the Hypertext Transfer Protocol (HTTP). For instance, if you look at the URL of the web page you are on at the moment, you could notice the prefix "http://", which tells the browser what protocol to use to execute the particular URL request.

Each web page's contents are usually presented in HTML or XHTML format, which allows for the information to be easily structured and then quickly read by the client's web browser. With the help of CSS (Cascading Style Sheets), designers can precisely control the web page's look and feel, as far as layout, typographic elements, color scheme and navigation are concerned. CSS instructions can be either embedded within the HTML web page (valid for that particular page) or can be included in a separate external file (valid for the whole site).

An example of How to Create a Simple Web Page

```
<html>
<head>
<title>Test my first page</title>
<link href="/style.css" rel="stylesheet" type="text/css" />
</head>
<body bgcolor="green" text="white">
<h1>My first Web page</h1>
Hello ! This is my first web page!
</body>
</html>
```

It's good to keep in mind that the web doesn't look the same to everyone. Some people use **Netscape**, **Internet Explorer**, Mosaic or other browsers that support **graphics**. Others can only make a **connection** through lynx, which supports the text and the links, but no pictures. Many members of the internet **community** only have access to web information through E-mail. Your modem speed and the type of connection that you have also has an effect on the way you view the WWW.

Sometimes, some pages will even **crash** your web browser. Don't **assume** that you did something wrong; it could be that the page is coded to offer **features** that your browser doesn't support. Many web designers use features that only work for the Netscape or Microsoft Internet Explorer browser, or for a certain operating system, etc.. There is nothing wrong with your computer. Just get your browser going again and you might want to shy away from sites that crash you!

Vocabulary

1. resource   *n.* 资源
2. FTP (File Transfer Protocol)   *n.* 文件传输协议
3. downloading software   *v.* 下载软件
4. document   *n.* 公文，文件，文献
5. URL (Uniform Resource Locator)   *n.* 统一资源定位器，网址
6. http (Hypertext Transfer Protocol)   *n.* 超文本传输协议
7. Netscape   美国 Netscape 公司，以开发 Internet 浏览器闻名

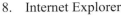

8. Internet Explorer		微软公司出品的 Web 浏览器
9. graphic		*n.* 图形
10. connection		*n.* 连接，联结，联系，关系，连接点
11. community		*n.* 社区，社会，团体，共有，共享
12. crash		*vt.&vi.* 使猛撞，使撞毁
13. assume		*vt.* 假设，臆断，假装
14. feature		*n.* 特征，特色，面貌，特写

Exercise 1

Match the words in the box with the descriptions below.

1. URL	2. web browser	3. FTP
4. Internet Explorer	5. Internet	6. http

a. It is the biggest network in the world.

b. Microsoft's web browser products

c. File Transfer Protocol

d. Hypertext Transfer Protocol

e. Uniform Resource Locator, is the address on the web that you are visiting.

f. It is an application software on your hard disk, which can find the text of the document (and other types of files) on the Internet and make them into the web page. Web page can contain graphics, audio and video, as well as text.

➢ Fast Reading

Text 1

Wireless Network

● **Wi-Fi**

Many coffee shops offer customers Internet access through a Wi-Fi connection. Wi-Fi is a wireless networking standard trademarked by the Wi-Fi Alliance. It refers to all networking equipment that is based on one of the IEEE 802.11 standards. Wi-Fi allows computers and other devices to connect to wireless routers and therefore other systems on the network. If the router is connected to the Internet, devices connected to the wireless access point may also have Internet access.

● **Bluetooth**

This wireless technology enables communication between bluetooth-compatible devices. It is used for short-range connections between desktop and laptop computers, PDAs (like the Palm Pilot or Handspring Visor), digital cameras, scanners, cellular phones, and printers.

Infrared once served the same purpose as bluetooth, but it had a number of drawbacks. For example, if there was an object placed between the two communicating devices, the transmission would be interrupted. (You may have noticed this limitation when using a television remote control.) Also, the infrared-based communication was slow and devices were often incompatible with each other.

Bluetooth takes care of all these limitations. Because the technology is based on radio waves, there can be objects or even walls placed between the communicating devices and the connection won't be disrupted. Also, bluetooth uses a standard 2.4 GHz frequency so that all bluetooth-enabled devices will be compatible with each other. The only drawback of Bluetooth is that, because of its high frequency, its range is limited to 30 feet. While this is easily enough for transferring data within the same room, if you are walking in your back yard and want to transfer the address book from your cell phone to your computer in your basement, you might be out of luck. However, the short range can be seen as a positive aspect as well, since it adds to the security of Bluetooth communication.

- **To create a wireless connection**

Click the Network icon in the task bar notification area. The Connections are available dialog box opens.

In Connections are available, ensure that Wireless Network Connection is expanded to reveal the list of available wireless networks. Click the name of the wireless network to which you want to connect. For example, if you want to connect to the Example wireless network, click Example. Click Connect. Depending on whether the wireless network is a secure or unsecured network:

If you are connecting to an unsecured network, the connection succeeds and you can begin to use the wireless network.

If you are connecting to a secured network where a security key is required, the Connect to a network dialog box opens. In Connect to a network, in Security key, type the security key, and then press ENTER.

Text 2

E-Commerce & M-commerce

E-commerce is simply business conducted online. The scope of electronic commerce includes a variety of activities related to the buying and selling of goods or services over the Internet. Some of the major activities often associated with e-commerce include the following.

- Virtual marketplaces that provide—e-tailing (online retailing)
- Internet-based marketing or advertising

- Online market research and surveys designed to gather product use, demographic, and other appropriate market research data
- Business-to-business (B2B), the computer-to-computer exchange of data and information between businesses
- Business-to-consumer (B2C), the electronic interactions between businesses and consumers via Internet server computers
- The use of a variety of technology-based approaches aiding interactions among businesses and between businesses and consumers (E-mail, fax, intranets, extranets, Internet telephoning, and so on)

E-Commerce Software and Web Site Integration

It is critical for you to ensure that your software, clearing service, and bank are linked to each other before taking your online payment system live.

Still another big problem that arises is that many merchants don't know exactly which company provides their commerce engine. Banks use different account setups for each commerce engine type. Therefore, an account might not work properly if you have it associated with the wrong commerce engine company when you set it up.

Security

From the very beginning, you'll want to check regularly that each of your e-commerce transactions is secure, Check your credit card statements carefully. If you use digital money, monitor your account balances.

Privacy

As Web sites collect more information about you, you're relinquishing privacy in return for convenience. Monitor how they can use your personal information and decide whether to submit this information in return for information and services.

Check the privacy policies at the e-commerce Web sites you shop. Do they sell your buying habits to other companies? What happens to this information if the company merges with another?

Trust

Whether the e-commerce Web site adheres to industry's privacy codes determines if you can trust it. Look for a TRUSTe logo, which means that the Web site follows certain standards. TRUSTe is a nonprofit organization that sets privacy standards and monitors its member Web sites for violations of privacy standards. You can learn more about TRUSTe at "www.truste.org".

> ## Supplementary Reading

HTML and CSS

HTML (the Hypertext Markup Language) and CSS (Cascading Style Sheets) are two of the

Unit 6 Surf the Network

core technologies for building Web pages. HTML provides thestructure of the page, CSS the (visual and aural) layout, for a variety of devices. Along with graphics and scripting, HTML and CSS are the basis of building Web pages and Web Applications. Learn more below about:

- **What is HTML?**

HTML is the language for describing the structure of Web pages. HTML gives authors the means to:

Publish online documents with headings, text, tables, lists, photos, etc.

Retrieve online information via hypertext links, at the click of a button.

Design forms for conducting transactions with remote services, for use in searching for information, making reservations, ordering products, etc.

Include spread-sheets, video clips, sound clips, and other applications directly in their documents.

With HTML, authors describe the structure of pages using markup. The elements of the language label pieces of content such as "paragraph" "list" "table" and so on.

- **What is XHTML?**

XHTML is a variant of HTML that uses the syntax of XML, the Extensible Markup Language. XHTML has all the same elements (for paragraphs, etc.) as the HTML variant, but the syntax is slightly different. Because XHTML is an XML application, you can use other XML tools with it (such as XSLT, a language for transforming XML content).

- **What is CSS?**

CSS is the language for describing the presentation of Web pages, including colors, layout, and fonts. It allows one to adapt the presentation to different types of devices, such as large screens, small screens, or printers. CSS is independent of HTML and can be used with any XML-based markup language. The separation of HTML from CSS makes it easier to maintain sites, share style sheets across pages, and tailor pages to different environments. This is referred to as the separation of structure (or: content) from presentation.

- **What is WebFonts?**

WebFonts is a technology that enables people to use fonts on demand over the Web without requiring installation in the operating system. W3C has experience in downloadable fonts through HTML, CSS2, and SVG. Until recently, downloadable fonts have not been common on the Web due to the lack of an interoperable font format. The WebFonts effort plans to address that through the creation of an industry-supported, open font format for the Web (called "WOFF").

Examples:

The following very simple example of a portion of an HTML document illustrates how to create a link within a paragraph. When rendered on the screen (or by a speech synthesizer), the link text will be "final report"; when somebody activates the link, the browser will retrieve the resource identified by "http://www.example.com/report":

```
<p class="moreinfo">For more information see the
<a href="http://www.example.com/report">final report</a>.</p>
```

The class attribute on the paragraph's start tag ("<p>") can be used, among other thing, to add style. For instance, to italicize the text of all paragraphs with a class of "moreinfo", one could write, in CSS:

```
p.moreinfo { font-style: italic }
```

By placing that rule in a separate file, the style may be shared by any number of HTML documents.

Section Two　Listening

Listening 1

About LAN system products

Tina Wu 是电子产品的零售商，她正在向 Frank 咨询有关新产品的情况。他们会怎么交谈呢？

Exercise 2

Listen to the dialogue and complete the notes outline.

Notes Outline：

1. Frank is introducing_____ to Tina Wu.
2. This adapter selects _____no matter where you are.
3. The card allows people to share the_____ and _____with the _____ in the company.
4. Tina will talk about it with _____ before she _____ Frank.

Exercise 3

Listen to the dialogue again and answer the following question.

1. Where does the dialogue most probably take place?
2. How many products are introduced?
3. How about the speed of the new wireless PC card?
4. Does Tina decide to buy the products?

Listening 2 Creating a new ADSL connection

Section Three Speaking

Exercise 4

What would you say in the following situations?

1. Imagine that you work at Lenovo Corp. and you are answering a phone call. Play your role according to the clues given in the brackets.（假设你在联想电信公司工作。现在你在接听电话。根据下面括号中的提示完成角色扮演对话。）

You:（电话铃声响起，你来接听，并说出 Lenovo Corp 的名称。）

Caller: Oh, hello. I'm calling because I have a problem with my home desktop computer. I can't log on to my Broadband LAN network. How can I create an Broadband LAN connection on my Windows XP?

You:（告诉对方在 Windows XP 中配置宽带连接的大致步骤。）

Caller: Well, I got it. Thanks a lot!

You:（表示客气，并祝对方愉快。）

2. A client called to ask how to create a wireless connection, please tell him the procedure briefly. (a. go to the Control panel and then Network Connections; b. create a new connection; c. select "Connect using a broadband connection"; d. enter the username and password.)

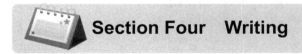

E-mail invitation 电子邮件邀请函

Sample 1:

From: eng@ichtf.com
To: angelafellini@yahoo.com
CC:
BCC:
Subject: Invitation to the 20th China Harbin International Economic and trade Fair
Date: May 15th, 2020

Dear Ms. Fellini,

Thank you for the interest you expressed in the China Harbin International Economic and Trade Fair (Harbin Trade Fair).

The 20th Harbin Trade Fair shall be held from June 15-19th, 2020 in Harbin International Conference Exhibition and Sports Center. It provides 3,000 international standard booths, the total exhibition space covers 86,000 square meters and 12 professional exhibition pavilions. During the fair, a series of business activities shall be held.

We hereby cordially invite you to take part in the 20th Harbin Trade Fair. Harbin Trade Fair shall be your ideal choice to acquire Chinese and international cooperation partners and shall provide qualified services up to your expectation.

I would be grateful if you could tell me the date on which you arrive. I do hope your schedule will allow you to accept our invitation.

Yours respectfully,

Organizing Committee of China Harbin International Economic and Trade Fair

Reply to the invitation (accepting) 回复邀请函（接受）

From: angelafellini@yahoo.com	
To: eng@ichtf.com	
CC:	
BCC:	
Subject: Re: Invitation to the 20th China Harbin International Economic and trade Fair	
Date: May 22nd, 2020	

Dear Sir/Madam

It was very kind of your to invite me to the China Harbin International Economic and Trade Fair (Harbin Trade Fair) on June15-19th, 2020.

I am very pleased to join the trade fair and will arrive in Harbin on Saturday, June 14th, 2020. My personal assistant, Judy Moore will accompany me during my visit. I believe we would enjoy our stay in the beautiful city, Harbin.

I will certainly look forward to seeing you at the trade fair.

Unit 6　Surf the Network

Yours faithfully

Angela Fellini
Marketing Manager

Reply to the invitation (declining) 回复邀请函（拒绝）

From: angelafellini@yahoo.com
To: eng@ichtf.com
CC:
BCC:
Subject: Re: Invitation to the 20th China Harbin International Economic and trade Fair
Date: May 22nd, 2020

Dear Sir/Madam

It was very kind of your to invite me to the China Harbin International Economic and Trade Fair (Harbin Trade Fair) on June15-19th, 2020.

Unfortunately, due to our new product launch taking place at the same time, I am unable to go to Harbin to take part in the trade fair. I hope that I will have an opportunity to join your next trade fair and do business with you in the future.

Best regards.

Yours faithfully

Angela Fellini
Marketing Manager

Sample 2:

From: catrinwilliams@126.com
To: lanjones@sohu.com
BCC:
Subject: invitation to visit Changyu IT company
Date: May 15th, 2020

Dear Mr. Jones

We would like to invite you to spend two or three days in Beijing to visit our company and let you have an overview of all our new models.

We would also arrange a sightseeing tour of the world-wide famous city and all your expenses will go to our accounts.

I do hope that you could accept our invitation.

Yours truly

Catrin Williams
Sales Manager

From: lanjones@sohu.com
To: catrinwilliams@126.com
CC:
BCC:
Subject: reply to your company visit invitation
Date: May16th, 2020

Dear Mr. Williams

I was delighted to receive your invitation to visit your company and would also be delighted to have a look at your products.

My business trip to the north starts from May 20th, 2020. I was wondering when would be possible for you to receive me.

I look forward to hearing from you.

Yours sincerely
Lan Jones

From: lanjones@sohu.com
To: catrinwilliams@126.com
CC:
BCC:
Subject: reply to your company visit invitation
Date:

Dear Mr. Williams

Thank you for your invitation. Unfortunately I am unable to visit your company due to a meeting I must attend in Shanghai at the same time.

Please do keep in touch and I would really like to go to visit your company some time in the future when my schedule allows.

Best wishes!

Truly yours
Lan Jones

Sample 3:

From: Jerryhardy@yeah.com
To: Simongreen@163.com
BCC:
Subject: invitation to a company event
Date: Sep 15th, 2020

Dear Mr. Green

I am writing to invite you as a VIP guest to join our company's new products presentation event on September 25th, 2020 in our computer superstore at 118 Nantong Avenue.

At the event, all our VIP guests will receive a notebook computer as a thank-you gift. I do hope that we can see you at our presentation.

Yours truly
Jerry Hardy
Sales Manager

Exercise 5

You work for Talent Computers Co. Ltd. You are asked to write an E-mail to invite your customer to attend your annual end-of-year dinner party. Also write two replies to the invitation. One is to accept the invitation and the other is to decline the invitation.

From:
To:
CC:
BCC:
Subject:
Date:

From:
To:
CC:
BCC:
Subject:
Date:

From:
To:
CC:
BCC:

Subject:
Date:

Transcript

Listening 1   About LAN system products

Tina Wu 是电子产品的零售商，她正在向 Frank 咨询有关新产品的情况。他们会怎么交谈呢？

A: Here we go, Ms. Wu. These are our new wireless PC Cards used for the wireless Intranet, an Internet access.

B: Please tell me something about the products, Frank.

A: Put the card in your notebook PC and no matter where you are, at home, at the office, at the airport or in other public access areas, this adapter automatically selects the best connection available, giving you constant access to E-mail and the Internet.

B: Can we access to our LAN using this card?

A: Certainly. You can share the information and equipment with the staff in your company.

B: It sounds nice. How about the speed of connection?

A: This PC card offers speeds up to nearly three times faster than other cards in homes and businesses.

B: That means we can save time and expenses.

A: That's right.

B: That's pretty impressive. Thank you for your introduction. Let me go back and talk to my boss, and then I'll contact you.

A: Thank you for your visit. I am looking forward to hearing from you.

Listening 2   Creating a new ADSL connection

A: Creative Software Corporation Customer Service. Hello, what can I do for you?

B: Oh, hello, I'm calling because I have a problem with my home desktop computer. I can't logon to my ADSL network. How can I create an ADSL connection on my Windows XP?

A: Oh, that's very easy if you have all the hardware in the right place. In Windows XP, go to

Control Panel, open Network Connections. Double click the Create New Connection icon, and now you will see a New Connection Wizard. In the second page, choose Connect to the Internet, and then choose Connect using a broadband connection that requires a username and password or the one that is always on. This will create a new ADSL connection for you.

B: OK, I see. I will try the best I can. Thanks so much.

A: Not at all. You're welcome to call Creative Software.

Unit 7 Selling Products

➢ Technical Reading

Here is part of an article about databases. First, read all the way through underling the basic features of a database.

Basic Features of Database

A **database** is a collection of data organized in a manner that allows **access**, **retrieval**, and use of that data. Database software, often called a **database management system (DBMS)**, allows users to create a computerized database; add, change, and **delete** the data; **sort** and **retrieve** the data; and create **forms** and reports from the data.

Information is entered on a database via **fields**. Each field holds a separate piece of information, and the fields are collected together into **records**. For example, a record about an employee might consist of several fields which give their name, address, telephone number, age, salary and length of employment with the company. Records are grouped tighter into **files** which hold large amounts of information. Fields can easily be updated: you can always change fields add new records or delete old ones. With the right database software, your are able to keep track of stock, sales, market trends, orders, invoices and many more detail that can make your company successful.

A database usually has more than one file. A **relational database** has links between its files. A relational database can avoid redundant data. It is easier to update the information and the data is consistent.

In a database:
- a **table** is another name for a file
- an **entity** is a subject about which information is stored in a table

- a **relationship** is a link between two entities
- an **attribute** is a property of an entity, for example Surname, Forename
- a **primary key** is an attribute used to ensure no two recodes are the same
- a **index** or **secondary key** is used when a table is often sorted into anther order
  Queries search through the database for required conditions. They can be written using
  - ➢ SQL (Structured Query Language)
  - ➢ QBE (Query By Example)

The **Database Administrator(DBA)** manages the database.

Vocabulary

1. database		*n.*	数据库
2. access		*vt.*	存取
3. retrieval		*n.*	检索
4. database management system(DBMS)		*n.*	数据库管理系统
5. delete		*vt.& vi.*	删除
6. sort		*vt.& vi.*	分类
7. retrieve		*vt.*	恢复
8. form		*n.*	窗体
9. field		*n.*	字段
10. record		*n.*	记录
11. file		*n.*	文件
12. relational database		*n.*	关系数据库
13. table		*n.*	表
14. entity		*n.*	实体
15. attribute		*n.*	属性
16. primary key		*n.*	主关键字
17. index		*n.*	索引
18. secondary key		*n.*	次要关键字
19. Database Administrator(DBA)		*n.*	数据库管理员

Exercise 1

Using the information in the text, complete these statements.

1. A database is used to _____.
2. Information is entered on a database via _____.
3. A _____ has links between its files.
4. An _____ is a subject about which information is stored in a table.
5. The _____ manages the database.

Unit 7　Selling Products

➢ Fast Reading

Text 1

The Structure of the Database Table

The first thing you do to set up a database table is to specify its structure by identifying the characteristics of each field in it. This structuring is done interactively, with the system prompting you to enter the field name, the filed type, and so on. For example, in the first row of the following figure the **field name** is COURSE ID; the **data type** is "text"; and the **field size**, or field length, is seven positions. The field names for the COURSE and STUDENT tables are listed at the top of each table in Figure 7.1 (COURSE ID, COURSE TITLE, TYPE, and so on). Content for a text field can be a single word or any **alphanumeric** (numbers, letters, and special characters) phrase. For number field types, you can specify the number of decimal positions that you wish to have displayed (none in the example because credit hours are whole numbers).

FIGURE 7.1 STRUCTURE OF THE EDUCATION DATABASE
This display shows the structure of the COURSE and STUDENT tables of the education database. The COURSE record(left) has four text fields and one number field. The STUDENT record has four text fields and one data/time field.

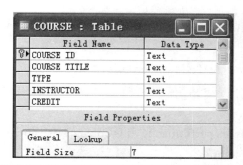

Figure 7.1

Vocabulary

1. field name　　　　　*n.* 字段名
2. data type　　　　　*n.* 数据类型
3. field size　　　　　*n.* 字段宽度
4. alphanumeric　　　*adj.* 文字数字的

Exercise 2

Read the information and answer the following questions.

1. What is the name of the two tables?
2. Which field is the key field of the COURSE table?
3. Which field provides a link to the COURSE table?

Text 2

DBMS & DBA

The **Database Management System (DBMS)** is the program that provides an interface between the database and user in order to make access to the data as simple as possible. It has several other functions:

1. **Data storage, retrieval and update.** Users to store, retrieve and up-date information as easily as possible. These users are not computer experts and do not need to be aware of the internal structure of the database or how to set it up.
2. **Creation and maintenance of the data dictionary.**
3. **Managing the facilities for sharing the database.** Many databases need a multi-access facility. Two or more people must be able to access a record simultaneously and to up-date it without a problem.
4. **Backup and recovery.** Information in the database must not be lost in the event of system failure.
5. **Security.** The DBMS must check passwords and allow appropriate privileges.

The security is the person responsible for maintaining the database. The administrator's tasks include the following:

1. The design of the database and monitoring its performance. If problems arise appropriate changes must be make to the database structure.
2. Keeping users informed of changes in the database structure that will affect them.
3. Maintenance of the data dictionary for the database.
4. Implementing access privileges-specifying what a user can access and /or change.
5. Protecting confidential information.
6. Allocating passwords to each user.
7. Providing training to users in how to access and use the database.
8. Ensuring adequate backup procedures exist to protect data.

Vocabulary

1. Database Management System (DBMS)   *n.* 数据库管理系统
2. data storage   *n.* 数据存储
3. retrieval   *n.* 检索
4. update   *n.* 更新
5. creation   *n.* 创造，新建
6. maintenance   *n.* 维护
7. data dictionary   *n.* 数据字典
8. facility   *n.* 设施，设备
9. sharing the database   *n.* 共享数据库
10. backup   *n.* 备份
11. recovery   *n.* 恢复
12. security   *n.* 安全

Exercise 3

Using the information in the text, complete these statements.

1. By_____, information in the database must not be lost in the event of system failure.

2. The DBMS must check_____ and allow appropriate_____.

➢ Supplementary Reading

SQL and PL/SQL

SQL has become the standard database language because it is flexible, powerful, and easy to learn. A few English-like commands such as SELECT, INSERT, UPDATE, and DELETE make it easy to manipulate the data stored in a relational database.

PL/SQL lets you use all the SQL data manipulation, cursor control, and transaction control commands, as well as all the SQL functions, operators, and pseudocolumns. This extensive SQL support lets you manipulate Oracle data flexibly and safely. Also, PL/SQL fully supports SQL datatypes, reducing the need to convert data passed between your applications and the database.

The PL/SQL language is tightly integrated with SQL. You do not have to translate between SQL and PL/SQL datatypes; a NUMBER or VARCHAR2 column in the database is stored in a NUMBER or VARCHAR2 variable in PL/SQL. This integration saves you both learning time and processing time. Special PL/SQL language features let you work with table columns and rows without specifying the datatypes, saving on maintenance work when the table definitions change.

Running a SQL query and processing the result set is as easy in PL/SQL as opening a text file and processing each line in popular scripting languages. Using PL/SQL to access metadata about database objects and handle database error conditions, you can write utility programs for database administration that are reliable and produce readable output about the success of each operation. Many database features, such as triggers and object types, make use of PL/SQL. You can write the bodies of triggers and methods for object types in PL/SQL.

PL/SQL supports both static and dynamic SQL. The syntax of static SQL statements is known at precompile time and the preparation of the static SQL occurs before runtime, where as the syntax of dynamic SQL statements is not known until runtime. Dynamic SQL is a programming technique that makes your applications more flexible and versatile. Your programs can build and process SQL data definition, data control, and session control statements at run time, without knowing details such as table names and WHERE clauses in advance.

Section Two　Listening

1. Please tell us the procedure to install the Windows XP operating system:
   (listening)
   The procedure to install the Windows XP operating system
   a. insert the Installation CD and reboot,
   b. format the primary partition,
   c. install
2. Please tell us how to install Sybase SQL Server well.
   (listening)

Section Three　Speaking

Exercise 4

Imagine that you work in Apple Inc., and now you are answering a phone call. Play your role according to the clues given in the brackets.（假设你在苹果公司工作。现在你正在接听电话。根据下面括号中给出的提示完成角色扮演对话。）

　　You:　（电话铃响了，您起来接听，并说出 Apple Inc.的名称。）

Unit 7　Selling Products

Lady: Hello, this is Catherine Lily. I am calling because I don't know how to install the Mac OS X 10. Could you please tell me the process step by step?
You:　（告诉对方安装 Mac OS X 10 的大致步骤。）
Lady: Well, I seem to understand the major steps now. Thank you very much!
You:　（表示客气，并祝对方心情愉快。）

Exercise 5

Imagine that you work at HuaWei Network. You are answering a phone call. Play your role according to the clues given in the brackets.（假设你在华为网络工作。现在你正在接听电话。根据下面括号中给出的提示完成角色扮演对话。）

You:　（电话铃声响起，你来接听，并说出 HuaWei Network 的名称。）
Caller: Hello, this is Philip Zhao. I am calling for assistance in integrating MySQL with Delphi. I am having some problems in making them work together.
You:　（告诉对方他应该去联系 Delphi 的技术支持，或者上网去找相关的教程或 FAQ。）
Caller: Well, if you say so, I'd try to talk to the Delphi staff first.

Exercise 6

1. Please tell us the procedure to remove installed programs.
2. Please tell us the procedure to use Microsoft Access to retrieve data from SQL Server database.

 Section Four　Writing

Writing E-mail of compliant 电子邮件（抱怨话题）

Sample 1:

From: caoyueyue@yeah.com
To: customerservice@hitech.com
CC:
BCC:
Subject: complaint about Lenovo laptop computer
Date: May 10th, 2021

To whom it may concern,

I am writing to complain about the Lenovo laptop computer I purchased from your computer store last Sunday, May 3rd, 2021 in the computer supermarket in Nantong Street.

The notebook kept switching off whenever I insert my mobile hard disk to it. Your maintenance engineer has already repaired it for me twice. I have no time to chase you to get it fixed again and again. This time I really want to have a replacement. Otherwise, I would like to return it and get the complete refund for it.

Please get in touch with me as soon as possible.

Yours,
Cao Yueyue

Sample 2:

From: pollyhan@yahoo.com
To: customerservice@qualitech.com
CC:
BCC:
Subject: complaint about the screen
Date: May 16th, 2021

To whom it may concern,

I bought a desktop computer with your promoted screen from your store yesterday, May 15th, 2021, unfortunately, the color produced by the screen looks not the same as you promised. There must be something wrong with it.

I would like you to send a maintenance man to have a look at it immediately otherwise I have to return it or get a refund.

Please contact me as quickly as you can.

Yours,
Polly Han

Sample 3:

From: headybrody@yahoo.com
To: customerservice@electroncorp.com
CC:
BCC:
Subject: complaint about the link of the internet
Date: May 23rd, 2021
To whom it may concern, I purchased a notebook computer from your store yesterday, May 22nd, 2021, unfortunately, the link with the Internet, which you set up for me is completely useless. It keeps going on and off for the whole afternoon today. I do not know what the matter is. Please send someone to repair it. I will wait for you in my office. Yours, Heady Brody

Exercise 7

You work for Talent Computers Co. Ltd. You are asked to write an E-mail to complain about the hardware you bought from Hitech Computers Co. Ltd.

From:
To: CC:
BCC:
Subject:
Date:

Transcript

1. Please tell us the procedure to install the Windows 10 operating system.

You should first clean up the primary partition, boot your computer with Windows 10 Installation Disc, and then format the primary partition C with the installer.

After that, you just follow the instructions, and do any necessary changes until it is all finished.

2. Please tell us how to install Sybase SQL Server well.

You should first configure your Operating System to suit the needs of your Sybase SQL Server, then you type "Su Sybase" to switch the user name to Sybase, and then you can install it by running Sy bload. After that you just follow the instructions.

Unit 8　With Customers

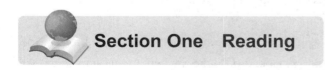
Section One　Reading

➤ Technical Reading

With support for secure access to corporate networks, directories, and Microsoft Exchange, iPad is ready to go to work.

iPad in Business

Mail, Contacts, and Calendar

To use iPad with your work accounts, you need to know the settings your organization requires. If you received your iPad from your organization, the settings and Apps you need might already be installed. If it's your own iPad, your system administrator may provide you with the settings for you to enter, or they may have you connect to a mobile device management server that installs the settings and apps you should have.

Organizational settings and accounts are typically in configuration profiles. You might be asked to install a configuration profile that was sent to you in an E-mail, or one that you need to download from a webpage. When you open the file, iPad asks for your permission to install the profile, and displays information about what it contains.

In most cases, when you install a configuration profile that sets up an account for you, some iPad settings can't be changed. For example, your organization might turn on Auto-Lock and require you to set a passcode in order to protect the information in the accounts you access.

You can see your profiles in Settings→General→Profiles. If you delete a profile, all of the settings and accounts associated with the profile are also removed, including any custom Apps your organization provided or had you download. If you need a passcode to remove a profile, contact your system administrator.

Network access

A VPN (Virtual Private Network) provides secure access over the Internet to private resources, such as your organization's network. You may need to install a VPN app from the App Store that configures your iPad to access a particular network. Contact your system administrator for information about any Apps and settings you need.

Apps

In addition to the built-in Apps and the ones you get from the App Store, your organization may want you to have certain other Apps. They might provide you with a pre-paid redemption code for the App Store. When you download an App using a redemption code, you own it, even though your organization purchased it for you.

Your organization can also purchase App Store App licenses that are assigned to you for a period of time, but which the organization retains. You'll be invited to participate in your organization's program in order to access these Apps. After you enroll with your iTunes account, you're prompted to install these Apps as they're assigned to you. You can also find them in your Purchased list in the App Store. An App you receive this way is removed if the organization assigns it to someone else.

Your organization might also develop custom Apps that aren't in the App Store. You install them from a webpage or, if your organization uses mobile device management, you receive a notification asking you to install them over the air. These Apps belong to your organization, and they may be removed or stop working if you delete a configuration profile or dissociate iPad from the mobile device management server.

Vocabulary

1. mail   n. 邮件
2. contact   n. 联系人
3. calendar   n. 日历
4. network access   n. 网络接入，网络访问
5. App (application 的缩略)   n. 应用程序，应用软件

Exercise 1

True or False

1. (   ) You install a configuration profile that does not set up an account for you.
2. (   ) To use iPad with your work, you do not need to know the any settings your organization requires.
3. (   ) If you delete a profile, all of the settings and accounts associated with the profile are not changed.
4. (   ) A VPN provides secure access over the LAN to private resources.
5. (   ) If your organization develops custom Apps, you may install them from a webpage.

Unit 8　With Customers

Fast Reading

Text 1

Create a PDF document in Mac OS X?

To create a **PDF** in OS X, you don't need Adobe software or a third party PDF creation tool. OS X provides a built-in way to create a PDF from any program that includes print functionality.

To save a document as a PDF, open the file and select File → Print. When the Print window opens, there is a drop-down menu in the left corner of the window. Click PDF button and select "Save as PDF…" as shown in the image (Figure 8.1) below.

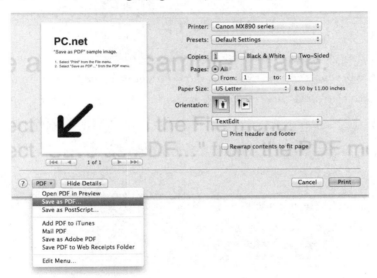

Figure 8.1

You can then save the PDF file in any folder, just like when you select "Save As…" from the File menu.

Since the "Save as PDF…" feature is available from the Print dialog box, it can be accessed from many different applications. If you need to share a document with a friend or colleague and he or she doesn't have the program you are using, one option is to save the document in the standard PDF format, then send the PDF.

The "Save as PDF…" feature is also useful when sharing sensitive data. PDFs are not as editable as other documents and cannot be modified at all in the free Adobe Reader program. Additionally, you can click the **Security Options** button when choosing where to save the PDF. This gives you the option to require a password to open or edit the document.

Vocabulary

1. PDF (Portable Document Format)   n. 便携式文件格式
2. Security Option   n. 安全选项

Exercise 2

True or False

1. (　) OS X can create a PDF from any program that includes print functionality.
2. (　) Selecting "Save as PDF…" may save a document as a PDF.
3. (　) PDF files are not as editable and modified in the free Adobe Reader program.
4. (　) Security Options button can help you to create a password for. pdf files.

Text 2

When dealing with customers, we must be as polite as possible. So when a customer is not satisfied with your products or services, you should apologize immediately and sincerely. In this activity, practice ways to make and respond to apologies.

A letter of complaint

From: john@hotmail.com
To: customerservice@telecomputers.com **Subject**: I demand an exchange for my laptop! **Date**: Thurs, Apr 23rd, 2020
Dear sir, Last week we purchased a laptop from your company. The problem is that the computer runs with much noise and its CPU fan doesn't work well, either. We have tried every means you have suggested, but nothing has worked. Since we are still within the **refund/exchange period**, we would like to ask for an exchange. Please reply to us at your earliest convenience. I will appreciate your help. Sincerely yours, Rose

Vocabulary

1. refund/exchange period   n. 退款/换货期

Exercise 3

Use the given sentences below, express your complaint.

1. I am writing to inform you that I am dissatisfied with your _____.
2. I understand you will give immediate attention to this matter _____.
3. I would like to have this matter settled by the end of _____.

A letter of response

From: customerservices@ telecomputers.com
To: lily@hotmail.com
Subject: Sincere apology for the behavior of our employee
Date: Mon, Apr 27, 2020
Dear Lily,   Please accept my apologies for the way you were treated on April 20th. Our company will never tolerate any rudeness to customers, and the person who was involved in this incident has been put on probation and has been sent to a special class on customer relations.   Thank you for bringing this matter to our attention. I sincerely hope this incident does not cause any damage to our business relationship.   Sincerely yours,   David Chen   Manager of Customer Services

Text 3

Read the following E-mail and complete the contact form according to what you've learned from the E-mail. You should type the answers in the blanks numbered 1 to 5.

> Linda,
>
> I've arrived at the hotel. The conference will start tomorrow and end next Monday. Send E-mails to me if anything props up during that period.
>
> I met someone when I was waiting for my flight in New York yesterday. I think she might be interested in our products. Could you note down her contact information and do the follow-up?
>
> Her name is Jane Song. She manages the Sales Department in NewLand Corporation,

which is based in Quincy, Massachusetts. They have interests in China. I've asked her to drop in when she's in Beijing next time. She wants to know about our NAS 800 series products. Send her a brochure on these, and offer her a special price if she asks. Her address is 115 West Squantum Street, Quincy, MA 02117.

Let me know if there is any problem.

Kind regards,

Patrick Lin
Manager of Sales Department
DreamTech

Contact Form	
Contact made by:	Patrick Lin
Place & Date:	New York, on 5 June
Contact Name:	Jane Song
Company:	(1)_____
Address:	115 West Squantum Street, Quincy, MA 02117
Department:	(2)_____
Position:	(3)_____
Interested in:	(4)_____
Actions:	(5)_____

➢ Supplementary Reading

Spreadsheet Program

A spreadsheet program is an application which is used to display data and to perform calculations on it. The word "spreadsheet" is also used to refer to the product created by this type of application—for example a table of information in words and figures, either on the screen, as a file, or as a printout.

In Excel, one of the most popular spreadsheet programs, the files used are called workbooks. These consist of worksheets, or "sheets", which are the electronic equivalent of large sheet of paper. The data is displayed in rows and columns, in a large table on the screen. The columns are usually labelled with letters of the alphabet (though sometimes with a number) at the top of the worksheet, and the rows are labelled with numbers on the left. The grey areas at the top of columns are called "column headings", and the corresponding areas to the left of the rows are called "row headings". Cells are the small boxes formed where rows and columns intersect. Every cell has an "address" which contains the labels of the row and column which intersect to form it.

Exercise 4

1. A spreadsheet program is used to _____.
    A. display data and perform calculations
    B. create the products
    C. label the column
2. The files used In Excel are called _____.
    A. worksheets
    B. workbooks
    C. sheets
3. The columns are usually labeled with _____.
    A. addresses
    B. numbers
    C. letters of the alphabet
4. Cells are the boxes where _____ intersect.
    A. rows and columns
    B. row headings and column headings
    C. letters and numbers

Exercise 5

Read the following paragraph. Choose the best word from A – G for each numbered blank. You should type in the blanks with the corresponding letters.

In Excel, one of the most popular (1)_____ programs, the files used are called workbooks. These consist of (2)_____, which are the electronic form of large sheets of paper. The data is displayed in rows and (3)_____, in a large table on the screen. Cells are the small boxes formed where rows and columns intersect. Every cell has a(n) (4)_____, which contains the labels of the row and column which intersect to form it, for example D5. Cells can contain numeric data, (5)_____, formulas or functions. Text is used for titles, or to describe the figures being displayed. Numeric data can be entered using the keyboard or can result from calculations.

A. address	B. text	C. workbooks	D. worksheets
E. spreadsheet	F. rows	G. columns	

Section Two Listening

Listening 1 Help customers to solve problems

Listening 2 Apologize to customers

Section Three Speaking

Exercise 6

1. Kevin promised to call a client Mr. Jack back yesterday, but he forgot to. What would he say in his telephone conversation with Mr. Jack?（Kevin 昨天许诺会给其客户杰克先生回电话，但是后来他忘了。在电话中他会和 Mr. Jack 说什么呢？）

   Kevin: Hello, is that ___(1)___ speaking?

   Mr. Jack: Yes, it is.

   Kevin: Oh, how are you, Mr. Jack? I am terribly sorry that I ___(2)___. I had a heavy work to do yesterday. Did you wait ___(3)___ yesterday?

   Mr. Jack: Oh, that's all right. Don't mention it. I just need you to fax me a new invoice.

   Kevin: I'll do that immediately. Sorry for ___(4)___.

   Mr. Jack: Don't worry about it.

2. A customer, Mr. White, is calling Jessica to make a reservation for a visiting support representative. The problem is the schedule for this week is full. What would they say in this situation? Filling the blanks below.（一个客户，怀特先生，打电话给 Jessica 预约客服代表。但是本周的日程已经满了。在这种情形下他们会说什么？）

   Jessica: DELL computers. ___(1)___?

   Mr. White: Well, I' am calling to make a reservation for a visiting support representative. There's something wrong with my network adapter.

   Jessica: All right. What date would be ___(2)___?

   Mr. White: What about this Friday?

   Jessica: ___(3)___, but the schedule for this week ___(4)___. Would next Monday ___(5)___ for you?

Unit 8　With Customers

Mr. White: OK. I will be at home next Monday. Thank you very much.
Jessica: You're ＿＿＿(6)＿＿＿. Have a nice day.

Exercise 7

- Try to apologize for something you can't do for a customer.
- Responding to other people's apology requires certain techniques as well. Humor and compliments work well. Try to imitate this pattern with your partner.

 Section Four　Writing

Reply to complaints 回复邮件（抱怨信函）

Sample 1:

From: customerservice@hitech.com
To: caoyueyue@yeah.com
CC:
BCC:
Subject: Re: complaint about Lenovo laptop computer
Date: May 12th, 2020
Dear Ms. Cao, On behalf of the Lenoro Company, I would like to express my deepest apologies for any inconvenience our laptop computer has caused you. As you requested, we will immediately send the replacement and a free 4GB-USB disk as a kind of compensation for the problem. We hope it will make you satisfied. Please contact me if you have any other special requirements. Yours, Liang Daxin After Sales Manager

Sample 2:

From: customerservice@qualitech.com
To: pollyhan@yahoo.com
CC:
BCC:
Subject: reply to the complaint about the screen
Date: May 16th, 2020

Dear Ms. Han,

Please accept our sincere apology for the inconvenience the screen of the desk-top computer has caused you and thank you for drawing our attention to this matter.

We would immediately send you a replacement and hope the new one will fit with your computer well.

Sincerely yours,
Leo Liang
Manager of Customer Services

Sample 3:

From: customerservice@electroncorp.com
To: headybrody@yahoo.com
CC:
BCC:
Subject: reply to the complaint about the internet link
Date: May 23rd, 2020

Dear Ms. Brody,

I do apologize for the inconvenience the bad link to the internet has caused you.

Charlie is already sent to help you solve the problem. Please tell him in details how the link breaks down. I am sure he will make it work.

Yours,
Polly Han

Exercise 8

Write a reply to the complaint about the recently bought hardware.

From:
To: CC:
BCC:
Subject:
Date:

Transcript

Listening 1

A: I am very sorry, but I don't think I can solve this problem right now.

B: What do you mean?

A: There must be something wrong with your hardware. I will get a hardware representative for you. I am really sorry for the inconvenience.

B: It's not a big deal. You have done a lot for me. Just get me the right person.

Listening 2

A: Sorry to have kept you waiting, Lucy. I was on a overseas call.

B: That's all right. You must be very busy.

A: Well, I hope you weren't bored while I was on the phone.

B: Not really. I was reading this new brochure of yours. I am very interested in your new product.

Unit 9　Solutions

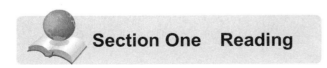
Section One　Reading

➢ Technical Reading

Presentation software lets you create highly stylized images for group presentations and kind, to create self-running slide shows for PC-based information displays at trade shows, for class lectures (offline or online), and any other situation that requires the presentation of organized, visual information. The software, such as Microsoft PowerPoint 2020, gives you a rich assortment of tools to help you create a variety of charts, graphs, and images and to help you make the presentation.

Presentation Software

Microsoft PowerPoint helps users to build visual **presentations** that help the create convey messages to an audience. After building your PowerPoint presentation, one easy way to display your presentation to an audience is by using a projector. LCD projectors allow you to display presentations from a computer in real time. Whether in the business world or education, the need to know how to use these tools is increasing. Even if you have never used these tools before, you can learn how to make PowerPoint and an LCD projector work together. Not all projectors are the same, but the same general principals apply. It is easiest to use a laptop when displaying a PowerPoint presentation due to the portability of the laptop. The type of laptop you use should have either an S-Video output or VGA output.

Things You'll Need
- Laptop computer
- PowerPoint Software
- LCD Projector

Instructions
- Find the PowerPoint presentation you want to present in your laptop and open it.
- Locate the "From Beginning" icon. Depending on which version of PowerPoint you are

using, the precise wording may vary; however, on PowerPoint 2020, you need to click on the "**Slide** Show" tab. In the "Start Slide Show" box you click on the first icon, "From Beginning".

- Place laptop on a flat surface near the LCD projector and check that your laptop's screensaver is disabled.
- Plug your LCD projector into a power outlet and turn on. The projector can take a few minutes to start up.
- Connect the VGA cable to the projector and then to the laptop. LCD projectors sometimes have more than one VGA cable connector. Search for the one that says "Line In" or "From Computer". Ensure the VGA cable is connected completely because if it is loose, it can cause display problems.
- Pull down the projection screen and center the image. Depending on the individual features of the projector, check to see that the display lens does not have a cover or sliding cover. Use the focus knob to sharpen the resolution of your presentation. You can also move the projector closer or farther away from the screen to adjust for size.
- Turn the LCD projector off and allow the cooling fan to stop.

Vocabulary

1. presentation   *n*. 报告，演讲
2. slide   *n*. 幻灯片

Exercise 1

True or False

1. ( ) Presentation software is necessary when displaying a PowerPoint presentation.
2. ( ) The reason that it is easiest to use a laptop when displaying a PowerPoint presentation is the portability of the laptop.
3. ( ) When a laptop displays a PowerPoint presentation, it does not need any S Video output or VGA output.
4. ( ) On PowerPoint 2020, you need to click on the "From Beginning" tab to display presention.
5. ( ) The article mainly introduces the requirements and instructions displaying presentation.

➤ Fast Reading

What if I need create slides for a presentation?

To increase the effectiveness of your class presentations and public speaking, presentation software can definitely help you. Presentation software helps you create and edit information that will appear in electronic sliders. And the information you include can be text, photos, art, tables, graphs, sound, animation, and even video.

As you create your slides, presentation software helps you define a consistent look and feel, change background colors, and add banners. You can also add link in your presentation and click on it, your presentation software will instantly start your Web browser software and go directly to the Web site for which you created a link.

Microsoft PowerPoint 2020 helps you prepare and present slides for presentations. PowerPoint has a variety of slide templates from which you can choose. You can work with entire presentation or with a single chart. Slides are easily rearranged by simply dragging a slide to a new position.

Microsoft PowerPoint 2020 has two types of templates: **design templates** and **content templates**. Design templates are predesigned formats and complementary color schemes with preselected background images you can apply to any content material (the outline) to give your slides a professional, customized appearance. A slide is one of the images to be displayed. Content templates go one step further and suggest content for specific subjects (for example, business plan, project overview, employee orientation, and many others).

Vocabulary

1. design template   n. 设计模板
2. content template   n. 内容模板

Exercise 2

1. Which types of templates are suggested content for specific subjects?
2. What can you do with design templates?

➤ Supplementary Reading

iPad mini User's Guide

iPad mini Features

The iPad Mini is Apple's small touch screen tablet. The iPad Mini maintains the same aspect ratio as the standard iPad, but reduces the screen size to 7.9 inches, which is smaller than the iPad as a whole. That's why it's 53% lighter. With iPad mini, you can:

- Store songs from your digital music collection, for listening on the go
- Listen to audio books purchased from the iTunes Music Store (you must have an Internet connection)
- Store and **synchronize** contact, calendar, and to-do list information from your computer
- Store text notes
- Set an alarm
- Play games, and more
- Photo and video shooting
- Edit document

To use iPad mini with a Windows PC, you must have: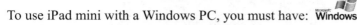

- A Windows PC with 500 MHz or higher processor speed
- Windows 7, or Windows 10 Home or Professional
- iTunes 12.0 or later
  To be sure you have the latest version of iTunes, go to "www.apple.com/itunes"
- **USB** 2.0 port, USB 3.0

Note: If your computer has a 4-pin FireWire port (see below), you can use it with iPod mini if you purchase an optional 6-pin-to-4-pin FireWire **adapter**. You can't charge iPod through your computer using a 4-pin FireWire port.

USB 2.0 port   USB 3.0 port

For more information about **compatible** FireWire and USB cards and iPod cables, go to "www.apple.Com/ipod".

Vocabulary

1. synchronize   *vt.* （使）同步
2. USB (Universal Serial Bus) port   *n.* 通用串行接口
3. adapter   *n.* 适配器
4. compatible   *adj.* 兼容的

Exercise 3

1. According to the passage, the iPod mini may not run well if the operating system is _____.

   A. Windows ME

   B. iPod software

   C. Windows PC with 500 MHz or higher processor speed

   D. iTunes 4.2 or later

2. To use iPad mini well, which of the following port will surely meet the requirement?

   A. Built-in 6-pin FireWire port.

   B. A built-in high-power USB 1.0 port.

   C. A 4-pin FireWire port.

   D. USB 2.0 or 3.0 port.

3. With iPod mini, you can do all but which of the following: _____.

   A. refer to a dictionary

   B. listen to audio books

   C. store text notes

   D. play games

Text 2

How to time your presentation

When you select Rehearse Timings, your Slide Show starts running, and PowerPoint starts timing it. The timing is displayed in the Rehearsal dialog box. When the Slide Show comes to an end, PowerPoint will give you the final running time.

Follow these steps to time your presentation:

1. Click the Slide Show menu, and then click Rehearse Timings. The Slide Show begins and a Rehearsal dialog box appears in the lower-right-hand corner of the screen.

2. Begin speaking and presenting your show.

3. If you want to repeat your rehearsal of a slide, click the Repeat button on the Rehearsal dialog box. The current slide repeats and the timing for it starts over.

4. Rehearse your presentation until it's finished. After you're done, a message box appears. It tells you the final running time and it asks you if you want to record the timings to use for viewing the presentation.

5. Click No. You are returned to the PowerPoint window.

Note: You click No because you are only timing the show. You would click Yes if you wanted to use the recorded timings to automate your presentation.

Exercise 4

1. When Rehearse Timings is selected, Slide Show _____.
   A. comes to an end
   B. starts running
   C. repeats
2. The final running time appears in _____.
   A. the Rehearsal dialog box
   B. the PowerPoint window
   C. the message box
3. You can begin speaking and presenting your show after clicking _____.
   A. the Rehearse Timings
   B. the Repeat button
   C. the Slide Show menu
4. If you want to time your rehearsal of a slide again, click _____ on the Rehearsal dialog box.
   A. the Rehearse Timings
   B. the Repeat button
   C. No
5. If you want to automate your presentation, click _____.
   A. Yes
   B. No
   C. the Repeat button

Section Two Listening

Dialogue

Listen to the following dialogue and answer the questions according to what you have heard.

Exercise 5

1. What is the name of the woman's company?
2. What has the man got?
3. Which data of department is he waiting for?

Dictation

Listen to the following short dialogue and fill in the blanks with the words you have heard.

A: Good morning. _____. May I help you?

B: May I speak to Mr. Liu? _____.

A: I'm sorry. You have reached _____. Mr. Liu is in the R&D Department. _____ to R&D.

B: Thanks. I appreciate that.

A: _____.

Section Three Speaking

Exercise 6

What would you say in the following situations?

1. Imagine that you work at Microsoft company. You are answering a phone call. Play your role according to the clues given in the brackets.（假设你在微软工作。你在接听电话。根据括号中的提示扮演角色。）

    You: （电话铃响起，您接听电话，并说出 Microsoft 的客户服务部。）

    Client: Hello, this is Sun Smith. I want to ask how to use PowerPoint 2007 to prepare a presentation. Can you tell me the process step by step?

    You: （告诉对方在 PowerPoint 中创建幻灯片的大致步骤，并提醒对方可以查阅用户手册 user's manual 或者联机帮助系统 FAQ online。）

    Client: Well, I seem to have got the idea. Thank you very much!

    You: （表示客气，并祝对方顺利。）

2. Imagine that you work at Microsoft Corp., and your client is giving you a call. Play your role according to the clues given in the brackets, and then take turns.（假设你在微软公司工作，而你的客户正在给你打电话。根据提示完成对话，然后交换角色。）

    You: （电话铃响起，接听电话，并说出 Microsoft 公司的名称。）

    Caller: （自我介绍。说明自己购买的 System Backup 9.0 不能运行，并提示系统缺少 .NET Framework。）

    You: （告诉对方下载、安装 .NET Framework 的方法。）

    Caller: （表明大致了解了安装方法。感谢对方。）

    You: （表示客气，并祝对方愉快。）

Section Four Writing

Faxes 传真

A fax is a common way that businesses and individuals communicate with each other. 传真是一种普通的企业或个人之间相互沟通的方式。It is fast and convenient as long as your fax machine is operating soundly. A written record of the letter sent is guaranteed as well. 它迅速、快捷，只需要保证传真机正常工作就可以确保信息的发送。

Sample 1:

To: Rowler Co.	**From:** DreamTech
Fax: 5536-2588	**Page:** 1
Phone: 6636-2258	**Date:** 05/06/2020
☐ **Urgent** ☐ **For Review**	☐ **Please Comment**
☐ **Please Reply**	☐ **Please Recycle**

Dear Sir/Madam

I am faxing to tell you that DreamTech is now opening the third chain store at 48, Xinyuan Avenue, Harbin.

Our store offers a new range of computer software packages for both personal and business use.

Enclosed is a list of the items we currently have in stock.

We hope that you will come and visit us soon.

Yours truly
Yifan Lin
Sales Manager

Sample 2:

To: View Co.	From: CompuTech
Fax: 5536-2596	Page: 1
Phone: 6636-2296	Date: 15/07/2020
□ Urgent □ For Review	□ Please Comment
□ Please Reply	□ Please Recycle

Dear Sir/Madam

I am sending a fax to tell you that CompuTech has manufactured a range of new models of mini notebook computers.

All our stores supply them from next Monday, 20th July, 2020. Large orders are welcome.

We hope that you will be interested.

Yours faithfully
Luyao Zhao
Sales Manager

Sample 3:

To: Dragon Co.	From: TopTech
Fax: 6615-3587	Page: 1
Phone: 6625-4587	Date: 12/08/2020
□ Urgent □ For Review	□ Please Comment
□ Please Reply	□ Please Recycle

Dear Mr. Barratt

The fax is sent to confirm that I have received the components we purchased from you. All parts are in good condition.

We would like to do more business with you in the future.

Best wishes!

Yours sincerely
Yuxin Gao
Purchasing Manager

Exercise 7

You work for DreamTech as a Sales Manager. Write a fax to your customer about the recently founded chain store.

Transcript

Dialogue

A: Hello. Creative Software.
B: Reynolds? This is Martin Gross calling.
A: Hi, Martin. What can I do for you?
B: Well, I've got the spreadsheets you need, but I am still waiting for the data from your accounting department.
A: Thanks so much for your help. Just bring me what you've got. That would be fine.

Dictation

A: Good morning. <u>GTT Software</u>. May I help you?
B: May I speak to Mr. Liu?<u> I need his help</u>.
A: I'm sorry. You have reached <u>the Marketing Department</u>. Mr. Liu is in the R&D Department. <u>Let me put you through</u> to R&D.
B: Thanks. I appreciate that.
A: <u>You are welcome</u>.

Unit 10   Computer Security

Section One   Reading

> Technical Reading

Kevin has been getting a number of questions about problems in computer security from customers recently. Let's read this document together, and learn along with Kevin.

Computer Viruses

The tern "computer **virus**" is a generic term for lots of different types of destructive software that spreads from file to file. A computer virus (virus) is software designed intentionally to cause annoyance or damage. There are two types of viruses, benign and malignant. The first type of virus displays a message or slows down your computer but doesn't destroy any information.

Malignant viruses, however, do damage to your computer system. Some will scramble or delete your files. Others shut your computer down, make your Word software act strangely, or damage the compact flash memory in your digital camera so that it won't store pictures any more.

One of the most prevalent types of computer viruses is the **macro** virus. Macro viruses are viruses that spread by binding themselves to software such an Word or Excel. If your computer is infected, and you send an infected file to someone else as an E-mail attachment, the recipient's computer will get the virus as soon as the attachment is opened. The virus will then make copies of itself and spread from file to file, destroying or changing the files in the process. You can also get infected if you download an infected file from the Internet or open a file on an infected disk.

A **worm** is a computer virus that spreads itself, not only from file to file, but from computer to computer via E-mail and other Internet traffic. A worm finds your E-mail address book and sends itself to the E-mail addresses in your list. One of the more famous worm viruses is called the "Love Bug". It arrives in your E-mail as an attachment to an E-mail. The subject of the E-mail is "I LOVE YOU"—a message that's hard to resist.

A virus can't do anything unless the virus instructions are **executed**. That usually means that you have to open the attachment to become infected because that's where the harmful code is. So be very careful about opening up an attachment if you're not sure what it is and where it came from. Also, be aware that some E-mail programs have the ability to execute macros (small blocks of code) and these may be executed when you open the E-mail itself, releasing the virus. Consult your E-mail program vendor for more information.

Vocabulary

1. virus   n. 病毒
2. macro   n. 宏，巨
3. worm   n. 蠕虫
4. execute   vt. 执行

Exercise 1

1. What is a computer virus?
2. How many type does a computer virus have?
3. Please list the removable media that can carry computer viruses.
4. What are the differences between a computer virus and a worm?

➢ Fast Reading

Text 1

How secure is your personal information?

Databases are an important part of e-commerce. Without them, businesses wouldn't be able to have interactive Web catalogs, process your credit card payment, or let you track your order. Businesses couldn't keep their inventories stocked or tell your about items that might interest you. Databases can store all of this data and produce information almost instantaneously.

Yet all this convenience comes with a price. It's the risk of people breaking into e-commerce Web sites. Crackers attack regularly. Most attempts fail, but some don't. And businesses don't often report the attacks because they don't want to damage their reputations. Consider these questions that relate to database security:

a. Would you work with or provide information to a business that a cracker had broken into? Would you improve its security after the break-in?

b. Who should be responsible for protecting data? Many businesses assume the software that they use can withstand an attack. They blame the software developers when a cracker finds a vulnerability.

c. Software developers blame businesses that don't monitor their daily "bug reports" on software systems. But reports discuss vulnerabilities and offer solutions. Should businesses be responsible for keeping their software current? Or should software developers notify businesses? Who should keep the system updated?

Many Web sites and organizations also inform people about software vulnerabilities and the latest cracks into e-commerce systems. Should you be responsible for monitoring the latest problems with database security? You can look at these Web sites to find out if a particular business war ever cracked.

Ultimately, who should be responsible for protecting valuable data that you provide to businesses?

Exercise 2

Read the following paragraph. Choose the best word from A – F for each numbered blank. You should type in the blanks with the corresponding letters.

Users have to enter a ___(1)___ to gain access to a network. You can download a lot of ___(2)___ or public domain programs from the net. Hundreds of ___(3)___ break into computer systems every year. A computer ___(4)___ can infect your files and corrupt your hard disk. A ___(5)___ is a device which allows limited access to an internal network from the Internet. Companies use ___(6)___ techniques to decode (or decipher) secret data.

| A. hackers | B. virus | C. firewall |
| D. password | E. files | F. decryption |

Text 2

Hackers!

Sept'70	John Draper, also known as Captain Crunch, discovers that the penny whistle offered in boxes of Cap'n Crunch breakfast cereal perfectly generates the 2600 cycles per second (Hz) signal that AT&T used to control its phone network at the time. He starts to make free calls.
Aug' 74	Kevin Mitnick, a legend among hackers, begins his career, hacking into banking networks and destroying data, altering credit reports of his enemies, and disconnecting the phone lines of celebrities. His most famous exploit —

	hacking into the North American Defense Command in Colorado Springs— inspired *War Games*, the 1983 movie.
Dec'87	IBM international network is paralyzed by hacker's Christmas message.
Jul'88	Union Bank of Switzerland "almost" loses $32 million to hacker-criminals. Nicholas Whitely is arrested in connection with virus propagation.
Oct'89	Fifteen-year-old hacker cracks US defense computer.
Nov'90	Hong Kong SAR introduces anti-hacking legislation.
Dec'92	Kevin Poulsen, known as "Dark Dante" on the networks, is charged with stealing tasking orders relating to an Air Force military exercise. He is accused of theft of US national secrets and faces up to 10 years in jail.
Feb'97	German Chaos Computer Club shows on TV the way to electronically obtain money form bank accounts using special program on the Web.
May'98	Computer criminals propagate a lot of viruses through the Internet.

Exercise 3

Read the above text then answer the following questions.

1. Which hacking case inspired the film War Games?
2. Why was Nicholas Whitely arrested in 1988?
3. How old was the hacker that cracked the US defense computer in October 1989?
4. Who was known as "Dark Dante" on the network? What was he accused of?
5. Which computer club showed on TV a way to attack band accounts?

➢ Supplementary Reading

Security and Privacy on the Internet

There are a lot of benefits from an open system like the Internet, but we bare also exposed to hackers who break into computer systems just for fun, as well as to steal information or propagate viruses. So how do you go about making online transactions secure?

Security on the web

The question of security is crucial when sending confidential information such as credit card numbers. For example, consider the process of buying a book on the web. You have to type your credit card number into an order form which passes from computer to computer on its way to the online bookstore. If one of the intermediary computers is infiltrated by hackers, your data can be copied. It is difficult to say how often this happens, but it's technically possible.

To avoid risks, you should set all security alerts to high on your web browser. Netscape Communicator and Internet Explorer display a lock when the web page is secure and allow you to disable or delete "cookies". A cookie is information placed on your hard disk by a Web site that you visit.

Privacy is the right to be left alone when you want to be, to have control over your own personal information, and to not be observed without your consent. A sniffer is software that sits on the Internet analyzing traffic.

If you use online bank services, make sure your bank uses digital certificates. A popular security standard is SET (Secure Electronic Transactions).

E-mail privacy

Similarly, as your E-mail message ravels across the net, it is copied temporarily on many computers in between. This means it can be read by unscrupulous people who illegally enter computer systems.

The only way to protect a message is to put it in as sort of "envelope", that is, to encode it with some form of encryption.

Spam is electronic junk mail, i.e., unsolicited mail, usually from commercial businesses attempting to sell your goods and services. Spoofing is forging the return address on an E-mail.

Network security

Private networks connected to the Internet can be attacked by intruders who attempt to take valuable information such as Social Security numbers, bank accounts or research and business reports.

To protect crucial data, companies hire security consultants who analyses the risks and provide security solutions. The most common methods of protection are passwords for access control, encryption and decryption systems and firewalls. A firewall is hardware and/or software that protects computers from intruders.

Virus protection

Computer viruses are software that was written with malicious intent to cause annoyance or damage. Viruses can enter a PC through files from disks, the Internet or bulletin board systems. If you want to protect your system, don't open E-mail attachments from a strangers and take care when downloading files form the Web (Plain text E-mail alone can't pass a virus.).

Remember also to update your anti-virus software as often as possible, since new viruses are being created all the time.

Exercise 4

Read the above text then answer these questions.

1. Why is security so important on the Internet?

Unit 10　Computer Security

2. What security features are offered by Netscape Communicator and Internet Explorer?
3. What security standard is used by most banks to make online transactions secure?
4. How can we protect and keep our E-mail private?
5. What methods are used by companies to make internal networks secure?
6. By which ways can a virus enter a computer system?
7. A _____ is hardware and/or software that protects computers from intruders.
8. _____ software prevents your being tracked while you're surfing.
9. The _____ tracks consumer fraud of all kinds.
10. Your _____ is the key to most information about you.

 Section Two　Listening

Section A: Short Conversations

In this section you will hear 5 short conversations. After each conversation, you will hear 1 question and choose the best answer from the 4 alternatives marked A, B, C and D. You will hear each conversation and question ONLY ONCE.

1. A. They've been working in the lab for two days.
   B. They can't find the piece of code.
   C. They've been working on computer for two hours.
   D. They can't fix the piece of code.

2. A. Tuesday.　　　　　　　　　B. Wednesday.
   C. Thursday.　　　　　　　　　D. Friday.

3. A. The woman will work in the International Department.
   B. The woman will work in the Domestic Department.
   C. The woman will jobless.
   D. The woman will take an job interview.

4. A. Linux.　　　　　　　　　　B. Windows 2003.
   C. Windows 2000.　　　　　　　D. Windows XP.

5. A. Bill will buy the laptop as soon as he gets the money.
   B. Bill's friends is buying the laptop for him.
   C. Bill can't afford a new laptop.
   D. Bill has already made the down payment on the laptop.

Section B: Monologue

One of your clients, David, called for a consultation on how to set up a firewall for the

Windows XP system, please offer your help.

（a. open Control Panel. b. configure the current network connection. c. enable the firewall.）

Section Three Speaking

Exercise 5

What would you say in the following situations?

1. Imagine that you work at Red Hat Linux company. You are answering a phone call. Play your role according to the clues given in the brackets.（假设你在 Red Hat Linux 公司工作，你在接听一个电话。根据括号中所给提示扮演角色。）

    You: （电话铃响起，您接听电话，并说出 Red Hat Linux 的技术支持部。）

    Client: Hello, this is Sun Smith. I want to ask how to enhance web security?

    You: （你可以购买防病毒和防火墙产品，安装在操作系统上。除此之外，在反病毒程序的基础上，再安装一个反间谍软件 anti-spyware 会是一个更好的选择。）

    Client: Oh, I see. Thanks a lot!

    You: （表示客气，并祝对方顺利。）

2. Imagine that you work at Microsoft, Inc. You are answering a phone call. Play your role according to the clues given in the brackets.（假设你在微软公司工作，现在你正在接听一个电话。根据括号中提供的线索扮演你的角色。）

    You: （电话铃响了，您接听电话，并说出 Microsoft 的名称。）

    Customer: Oh, hello. This is Samuel Smith speaking. I want to make some changes to my system configuration. Other users keep logging onto my computer with my user name. How can I prevent them from logging in?

    You: （告诉对方在 Windows 中设置账号密码的大致步骤。）

    Customer: It is done. Thank you so much.

    You: （表示客气，并祝对方愉快。）

Section Four Writing

Product Description 描述产品

Product description includes the following points: 描述产品包括以下几点：

Unit 10 Computer Security

1. Name of product 产品名称：model and brand 型号和品牌
2. Type of product 产品类型：electrical goods 电器, electronic products 电子产品, food, etc.食品等
3. Performance or functions 性能或功能：How well does it work? 性能如何？And what functions does it have? 有什么功能？
4. Features 特点：extra special functions 特殊功能；attraction（有什么）吸引（消费者的特点）
5. Reliability 可靠性：Is it likely to go wrong easily in the first year or the first two years?
6. Durability 耐用性/持久性：Does it last long?
7. Delivery, Guarantees and After-sales Service 送货、保修期和售后服务：Is it delivered to your place? How long is it guaranteed?

Example 1:

Describing a computer 描述一台电脑

Name: a computer (A particular model name is better.)

Type of product: Electrical goods.

Performance or function: It is used to store and retrieve information. Its function is to help the user keep a record of anything. It has different ways to input information. For example, by using a keyboard, a floppy disk, the modem over a telephone link or from another computer linked with this one. It also has different ways to output information, such as on a screen called a VDU (Visual Display Unit), a printer, a floppy disk or another computer.

Features: It can have high resolution colour monitor, come complete with word-processing and other business software.

Reliability: The excellent hardware enhances its reliability.

Durability: It will last for ever.

Delivery: Free delivery service is offered.

Guarantee: 2 year express exchange guarantee.

After-sales Service: Our repairman is ready for any necessary training and solving all problems.

Example 2:

Detailed Product Description 产品细节描述

Huawei MateBook X Pro

Details

Operation system: Windows 10 64 bit home Chinese version

Display: 13.9 inch

CPU: Intel Core i7-1165g7

Memory: 16GB

HDD: 512GB SSD 1TB SSD

Built-in Camera: 1 megapixel camera

Sound Effect: Dolby atmos panoramic sound effects

DC-In: 1 port

Earphone: 1 port

Micro Phone: 1 port

RJ45 Port: 1 port

VGA Port: 1 External D-15 output

USB Port: 1*USB2.0 + 1*USB3.0, USB type-C interface

Battery: 100v-240v 65W adaptive AC power adapter

Dimension: 355mm×272mm×67mm

Weight: 1kg (including) - 1.5kg (excluding)

(Reference: https://detail.zol.com.cn/1311/1310254/param.shtml)

Sample 3:
iMac LCD screen description (advertisement style) 苹果液晶显示器描述（广告形式）

Beauty by the inch

Photos, movies, games, videos, and applications with palettes. Anything you see on the 27-inch glossy widescreen display will be a pixel-perfect experience. iMac features a flat-panel LCD screen with 5120×2880 (27-inch), giving you vivid colors and breathtaking high-definition clarity. And there's nothing quite like seeing your life's events on the big screen, thanks to the new iLife '09 built into every iMac. Even more eye opening, the 27-inch iMac starts at just $2171.

Exercise 6

Fill in the gaps. Use the following words:

> video game; 4MB; performance; notebook; Computer

Apple MacBook Pro 17" Laptop Computer

The Apple MacBookPro Laptop ___(1)___ is all about ultimate speed,___(2)___, connectivity, and mobility. Built on the revolutionary Intel Core 2 Duo which packs the power of two processor cores (in this case 2.4GHz) inside a single chip, it provides ___(3)___ of Smart Cache, L2 cache that can be shared between the cores as needed. Whether you're creating 3D animation, editing photos from the day's shoot, or playing your favorite, the MacBook Pro delivers serious processing power with the AMD Radeon Pro 5500M graphics controller. It also delivers up to 16GB of DDR4 2666MHz ___(4)___ memory with a faster, 800MHz frontside bus and support for hard drives up to 1TB. Amazingly enough, all this power comes in a package only one inch thick and a mere 2kg, making the MacBook Pro model among the thinnest, lightest ___(5)___ computers in its display class.

Transcript

Listening 1

Please go to the Control Panel, enter "Network Connections" by double clicking it. Then right click on your current connection, such as your LAN, ISDN or ADSL connection, and select Propertied. In the Advanced tab, enable the Internet Connection Firewall by clicking the "On" radio button. That's all.

Unit 11 The Development Environment

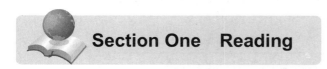

Section One Reading

➤ Technical Reading

Have you ever wondered how computer programs work? Have you ever wanted to learn how to write your own computer programs? Whether you are 14 years old and hoping to learn how to write your first game, or you are 70 years old and have been curious about computer programming for 20 years, this article is for you.

Programming Language

A programming language is a formal language designed to communicate instructions to a machine, particularly a computer. Programming languages can be used to create programs that control the behavior of a machine and/or to express algorithms precisely.

The description of a programming language is usually split into the two components of syntax (form) and semantics (meaning). Unfortunately, computers cannot understand ordinary spoken English or any other natural language. The only language they can understand directly is called **machine code**. This consists of the 1s and 0s (binary codes) that are processed by the CPU.

However, machine code as a means of communication is very difficult to write. For this reason, we use symbolic languages that are easier to understand. Then, by using a special program, these languages can be translated into machine code. For example, the so called **assembly languages** use abbreviations such as ADD, SUB, MPY to represent instructions.

Basic languages, where the program is similar to the machine code version, are known as **low-level languages**. In these languages, each instruction is equivalent to a single machine code instruction, and the program is converted into machine code by a special program called an assembler. These languages are still quite complex and restricted to particular machines.

To make the programs easier to write and to overcome the problem of intercommunication between different types of machines, higher-level languages were designed such as BASIC, COBOL, FORTRAN or Pascal. These are all problem-oriented rather than machine-oriented. Programs written in one of these languages (known as **source programs**) are converted into a lower-level language by means of a **compiler** (generating the **object program**). On compilation, each statement in a **high-level language** is generally translated into many machine code instructions.

People communicate instructions to the computer in symbolic languages and the easier this communication can be made the wider the application of computers will be. Scientists are already working on Artificial Intelligence and the next generation of computers may be able to understand human languages.

Instructions are written in a high-level language (e.g. Pascal, BASIC, COBOL, C). This is known as the source program.

Compiler
Compilers translate the original code into a lower-level language or machine code so that the CPU can understand it.

Instructions are compiled and packaged into a program. The software is ready to run on the computer.

Vocabulary

1. machine code   *n.* 机器代码
2. assembly language   *n.* 汇编语言
3. low-level language   *n.* 低级语言
4. source program   *n.* 源程序
5. compiler   *n.* 编译程序
6. object program   *n.* 目标程序
7. high-level language   *n.* 高级语言

Unit 11   The Development Environment

Exercise 1

Read the text and find answers to these questions.
1. Do computers understand human languages?
2. What are the differences between low-level and high-level languages?
3. What is an assembler?
4. What is the function of compilers?
5. What do you understand by the terms source program and object program?

➢ **Fast Reading**

Text 1

Your First Program

Your first program will be short and sweet. It is going to create a drawing area and draw a diagonal line across it. To create this program you will need to:
1. Open Notepad and type in (or cut and paste) the program
2. Save the program
3. Compile the program with the Java compiler to create a Java Applet
4. Fix any problems
5. Create an HTML web page to "hold" the Java Applet you created
6. Run the Java Applet

Here is the program we will use for this demonstration:

```
import java.awt.Graphics;
public class FirstApplet extends java.applet.Applet
{
    public void paint(Graphics g)
    {
        g.drawLine(0, 0, 200, 200);
    }
}
```

Step 1 - Type in the program

Create a new directory to hold your program. Open up Notepad (or any other text editor that can create TXT files). Type or cut and paste the program into the Notepad window. This is important: When you type the program in, case matters. That means that you must type the uppercase and lowercase characters exactly as they appear in the program. Review the programmer's creed above. If you do not type it EXACTLY as shown, it is not going to work.

Step 2 - Save the file

Save the file to the filename FirstApplet.java in the directory that you created in step 1. Case matters in the filename. Make sure the "F" and "A" are uppercase and all other characters are lowercase, as shown.

Step 3 - Compile the program

Open an MS-DOS window. Change directory ("cd") to the directory containing FirstApplet.java. Type:

javac FirstApplet.java

Case matters! Either it will work, in which case nothing will be printed to the window, or there will be errors. If there are no errors, a file named FirstApplet.class will be created in the directory right next to FirstApplet.java.

(Make sure that the file is saved to the name FirstApplet.java and not FirstApplet.java.txt. This is most easily done by typing dir in the MS-DOS window and looking at the file name. If it has a. txt extension, remove it by renaming the file. Or run the Windows Explorer and select Options in the View menu. Make sure that the "Hide MD-DOS File Extensions for file types that are registered" box is NOT checked, and then look at the filename with the explorer. Change it if necessary.)

Step 4 - Fix any problems

If there are errors, fix them. Compare your program to the program above and get them to match exactly. Keep recompiling until you see no errors. If javac seems to not be working, look back at the previous section and fix your installation.

Step 5 - Create an HTML Page

Create an HTML page to hold the applet. Open another Notepad window. Type into it the following:

<html>

<body>

<applet code=FirstApplet.class width=200 height=200>

</applet>

</body>

</html>

Save this file in the same directory with the name applet.htm.

Step 6 - Run the Applet

In your MS-DOS window, type:

appletviewer applet.htm

You should see a diagonal line running from the upper left corner to the lower right corner:

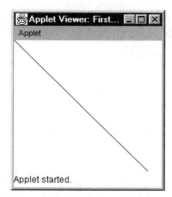

Pull the applet viewer a little bigger to see the whole line. You should also be able to load the HTML page into any modern browser like Netscape Navigator or Microsoft Internet Explorer and see approximately the same thing.

You have successfully created your first program!

Java: Platform-independent language used for Web-based applications.

HTML (HyperText Markup Language): The language used to compose and format most of the content found on the Internet.

Applet: A small program sent over the Internet or an intranet that is interpreted and executed by Internet browser software.

Exercise 2

Study this table about Java and answer the questions below:

Language	Date	Characteristics	Uses
Java Invented by Sun Microsystems.	1995	Cross-platform language that can run on any machine. Small Java Programs, called "applets", let you watch animated characters, play music and interact with information.	Designed to create Internet applications. When you see a web page containing Java links, a Java program is executed automatically.

1. Who invented Java?
2. When was Java developed?
3. Can Java run on any computer (Mac, PC or UNIX workstation)?
4. What are Java's small programs called? What can you do with them?

(For more examples of Java, visit: www.gamelan.com.)

Text 2

Look carefully at the job advertisements and discuss with another student what personal qualities and professional abilities you would need for each job.

Job One

A pro-active role in a legal environment. The role will be primarily to provide support to around 100 users over two sites supporting a newly introduced application and also other desktop applications.

Applicants will be part of a small team, working on a daily basis with desktop users therefore good communication, documenting and reporting skills are essential.

Experience: Minimum 12 months experience supporting a corporate environment.

Primary skills: Windows XP in and Active Directory environment, MS Office 2002, Networked Printers (HP) Secondary skills (Good Working Knowledge): Windows 2000 server and above, MS Exchange, LAN/WAN protocols, Spam and Anti Virus systems, Remote Access, VPN, Backup procedures Advantageous: MCP/CCNA Certification, SQL Server, Previously supported a legal environment.

Job Two

London based IT service provider is seeking Functional and Automated Software Test engineers with 1-3 years of experience. A testing engineer is capable of supporting a client in test specification, execution and reporting needs. Tasks include creating test scripts using industry standard test specification methods as well as test execution during different phases of the project life cycle and analysis of and reporting on findings. ISEB Foundation Certificate in Software Testing is preferred.

Sogeti's worldwide service offering regarding testing, provides customers with insight into the quality of their software and the risks associated with insufficient quality. Sogeti methodologies like TMap® (Test Management Approach) and TPI® (Test Process Improvement) are used by companies all over the world. Sogeti employs over 1500 testing and QA professionals worldwide and seeks to significantly increase its testing presence in the UK.

Essential Skills:
- Functional Software Testing
- Automated Software Testing (multiple tools)
- ISEB Foundation Certificate in Software Testing

Desirable Skills:
- Defect Tracking and Management
- Performance testing
- Mercury toolset (WinRunner, QuickTestPro, LoadRunner, TestDirector)

Job Three

Responsible for specifications, design and implementation for custom reports and new features in Active Server Pages and Web development.

Yardi Systems, Inc. is a great place to work. We have terrific employees, a pleasant and casual working environment, competitive salaries, and an unbeatable benefits package. Our corporate culture stresses integrity, respect, trust, responsibility, and fun. We look for professional, enthusiastic, and self-motivated team players with a desire to learn and the ability to work in a fast paced environment. In today's economy, there is some security in knowing that our clients have been using our real estate management software since 1982.

Requirements

Strong technical knowledge in software development methodologies, design and implementation. Thorough technical knowledge in coding VB.net, Oracle and MS-SQL programs. Bachelor's degree in Computer Science or equivalent experience. (0-2 years exp preferred)

Exercise 3

1. What position is the ad of Job One about?
    A. Computer programmer.
    B. Test engineer.
    C. Technical support.

2. If you are a new graduate from a technical university, which job may best suit you?
    A. Job One.
    B. Job Two.
    C. Job Three.

3. Which position requires good communication skills as well as essential technical knowledge?
    A. Job One.
    B. Job Two.
    C. Job Three.

Exercise 4

1. Tick the most important qualities in the list. Then add some more of your own.

logical reasoning	☐	ability to draw well	☐
ability to lead a team	☐	efficiency	☐
imagination	☐	being good with figures	☐
patience and tenacity	☐	self-discipline	☐
physical fitness	☐	willingness to take on responsibility	☐

2. Would you like to apply for one of these jobs? Why?

3. Study the personal profile of Charles Graham. Which it the most suitable job for him?

Charles Graham

- 28 years old. Married
- Education: LSBU London
- In-depth knowledge of Apple Macintosh equipment
- Course in graphic design and page-layout applications form Highland Art School
- Proficient in Adobe PageMaker and Super Paint
- Diploma in word processing. Wide experience in MS Word, and WordPerfect
- Present job: Computer operator for PromoPrint, a company specializing in publishing catalogues and promotional material

➢ Supplementary Reading

What is Java?

Java is a programming language and computing platform first released by Sun Microsystems in 1995. There are lots of applications and websites that will not work unless you have Java installed, and more are created every day. Java is fast, secure, and reliable. From laptops to datacenters, game consoles to scientific supercomputers, cell phones to the Internet, Java is everywhere.

Java allows you to play online games, chat with people around the world, calculate your mortgage interest, and view images in 3D, just to name a few. It's also integral to the intranet applications and other e-business solutions that are the foundation of corporate computing. Java is free to download. We can get the latest version at java.com. The latest Java version contains important enhancements to improve performance, stability and security of the Java applications that run on your machine. Installing this free update will ensure that your Java applications continue to run safely and efficiently.

The Java Runtime Environment (JRE) is what you get when you download Java software. The JRE consists of the Java Virtual Machine (JVM), Java platform core classes, and supporting Java platform libraries. The JRE is the runtime portion of Java software, which is all you need to run it in your Web browser. When you download Java software, you only get what you need—no spyware, and no viruses. The Java Virtual Machine (JVM) is only one aspect of Java software that is involved in web interaction. The Java Virtual Machine is built right into your Java software download, and helps run Java applications.

The Java Plug-in software is a component of the Java Runtime Environment (JRE). The JRE allows applets written in the Java programming language to run inside various browsers. The Java Plug-in software is not a standalone program and cannot be installed separately.

Section Two Listening

Listening 1 Java Running Environment Java 运行环境

Listening 2 Specifying the Development Environment 指定开发环境

Section Three Speaking

Exercise 5

What would you say in the following situations?

1. Imagine that you work for SunTech Software Outsourcing Company, and your partner is giving you a call. Play your role according to the clues given in the brackets, and then take turns. （假设你在 SunTech 软件外包公司工作，而你的搭档正在给你打电话。根据提示完成对话，然后交换角色。）

You: （电话铃响起，接听电话，并说出 SunTech 公司的名称。）
Caller: （自我介绍。说明自己公司委托对方开发的信息加密系统要求使用 C# 而不是 VB 开发。）
You: （表示一定将对方的要求转述给技术部门的主管，并让他与对方联系。）
Caller: （感谢对方。）
You: （表示客气，并祝对方愉快。）

2. How to run Scheduler Home Edition on Windows XP?（如何在 Windows XP 上使用 Scheduler 家庭版？）

Section Four　Writing

Writing a CV & a Covering letter 简历和求职信的写作

A CV (Curriculum Vitae) is a personal marketing tool to promote oneself in finding a job. 简历是人们找工作时推销自己的一个工具。Emphasize one's skills and experiences to the potential employer with their strong points to make their application stand out. 向招聘单位说明自己突出的技能、经验以及强于他人之处。The aim is to get the opportunity to be interviewed by the employer. 目的是得到面试的机会。

A CV should include the following points: 简历应该包括以下内容：

1. Personal information 个人信息（name 姓名, address 住址, E-mail 电子邮箱地址, contact numbers, etc. 联系电话等。Don't write too much 不要写太多）
2. Career target or job objective 求职意向（list the positions you want to apply for to show your expectations 列举自己期望应聘的职位）
3. Profile 简介（brief your abilities, strengths, etc. use expressive adjectives + noun phrases 简单地介绍自己的能力、优势等，使用形容词+名词的表达方法）
4. Education 教育（背景）（include the school, the degree or diploma, the major and the time, etc. 写明所读学校、所获得的毕业证和学位证、所学专业、受教育的时间等信息）
5. Work Experience 工作经验（write the name of employer, job title, duties/responsibilities and the time, etc. 写明工作单位、职位、职责、工作时间等。）
6. Other skills 其他技能（computer skills and foreign language skills 计算机和语言技能）
7. Qualifications/Certifications 资格证书、能力证书（list all degrees, diplomas, certificates 列举所有的学位证书、毕业证书、能力证书等）
8. Hobbies 兴趣爱好（mention the good ones which help enhance your image 列举能增强个人形象的、好的兴趣爱好）

Layout of a CV 个人求职简历的格式

CV

Personal information (name, address, E-mail, contact numbers, etc. don't write too much)

Career target or job objective (list the positions you want to apply for to show your expectations)

Profile (brief your abilities, strengths, etc. use expressive adjectives + noun phrases)

Education (include the time, the degree or diploma, the major and the school, relevant coursework, etc.)

Work Experience (write the time, job title, the name of employer, duties/responsibilities, etc)

Other skills (computer skills and foreign language skills)

Qualifications/Certifications (list all degrees, diplomas, certificates)

Hobbies (mention the good ones which help enhance your image)

Sample 1:

CV

Wang Jinsheng

Personal Information

Address: Pobox 5 HNU Star College, 1st Star Road, Hulanlimin Development Zone, Harbin

Contact Number: (mobile)130××××8858; (home) 0451-8810××××

E-mail: wangjsh@gmail.com

Career Target/Job Objective

Software Programmer/Software Engineer/A position in software development company

Profile

Talented software programmer with BBA degree, strong educational background in programming, and experience using cutting-edge development tools. Articulate and professional communication skills, including formal presentations and technical document. Productive in both team-based and self-managed projects; dedicated to maintaining up-to-date industry knowledge and IT skills.

Education

Sep 2004-Jul 2008 Bachelor's Degree in Computer Science at Heilongjiang University

Relevant coursework

Software Design; Database, System Analysis, Project Management, Web Site Design and Development, Project Teamwork and Communication, Technical & End User Documentation

Social practice projects involved:

Software Engineering — a member of programming project team working on the development of software for actual implementation with Harbin College Resources Centre

Work Experience

Aug 2008 Software Developer at APC Computer Software Co. Ltd.

Duties:

Worked with small team of developers to work out ideas for redesigning the existing application software used in the company

Computer Skills

Languages: C, C++, Java Script, UNIX Shell, FORTRAN, HTML

Operating Systems: UNIX, Windows XP/Windows 7/Windows 10

Software/Applications: MS Visual C++, Visual Basic, Showcase, MS Access, MS Excel, MS Internet Explorer, Netscape, Adobe Photoshop, Adobe Acrobat, Microsoft Office,

Microsoft Project
Hardware: HP-UX, Sun Sparc, PC, Macintosh, SGI-ONYX, SGI Crimson
Networking: TCP/IP, Microsoft LAN Manager, Novell Netware
Qualifications/Certifications: CET4/6, NIT, Undergraduate Diploma, Bachelor's Degree of Computer Science, CCIE, MCSE, SCJP, DB2 Database Administrator
Hobbies
Surfing the Internet, Reading, Swimming

Sample 2:

CV
Ma Shaolin

Personal Information
Address: 61st Jianguo St, Daoli District, Harbin, China
Contact Number: (mobile)130××××9959; (home) 0451-8460××××
E-mail: mashl@gmail.com
Career Target/Job Objective
System Analyst/Computer Consultant/Lab Attendant/System Administrator/Office Assistant
Profile
Gifted software developer with a Bachelor's degree of Business Administration in Computer Science. Excellent educational background in programming, and experience using different development tools. Good communication skills, including formal presentations and technical document. Productive in both team-based and self-managed projects; dedicated to maintaining up-to-date industry knowledge and IT skills.
Education
Sep 2005-Jul 2009 Bachelor's Degree in Computer Science at HNU Star College
Relevant coursework:
Software Design; Database, System Analysis, Project Management, Web Site Design and Development, Project Teamwork and Communication, Technical & End User Documentation
Social practice projects involved:
Software Development — worked with a small team of developers to brainstorm and implement ideas for shipping software with electricity supply industry
Work Experience
May 2009 Software Developer at Acer Computer Software Co. Ltd.

Duties:
Helped solve customers' problems with the software programmed by the company

Computer Skills

Languages: C, C++, Java Scrip, JSP, ASP, SQL

Operating Systems: UNIX, Windows XP/Windows 7/Windows 10

Software/Applications: MS Visual C++, Visual Basic, SQL Server 7.0/2000, MS Exchange 5.5/2000, MS Internet Security Acceleration Server

Hardware: HP-UX, Sun Sparc, PC, Macintosh, SGI-Octane

Networking: TCP/IP, Microsoft LAN Manager, Novell Netware

Qualifications/Certifications: BEC P/V/H, NIT, Undergraduate Diploma, Bachelor's Degree of Computer Science, CCIE, MCSE, SCJP, DB2 Database Administrator

Hobbies
Reading, Writing, Travelling

Sample 3:

CV

Zhang Jing

Personal Information

Address: 51st Nanzhi Rd, Daowai District, Harbin, China

Contact Number: (mobile)135××××1357; (home) 0451-5769××××

E-mail: zhangjing@gmail.com

Career Target/Job Objective

Computer Consultant/Lab Attendant/Office Assistant

Profile

Gifted software user with a Bachelor's degree of Business Administration in Computer Science. Excellent educational background in using different software. Good communication skills, including formal presentations and technical document. Productive and efficient as a team member or an individual; dedicated to learning to use new software.

Education

Sep 2004-Jul 2008 Bachelor's Degree in Computer Science at Harbin Commercial University

Relevant coursework:

Microsoft office; Computer Language C, Database, Web Site Design and Development, Project Teamwork and Communication

Social practice involved:

worked as an office clerk for LITC, a computer software development company

Work Experience

October-December 2008 Software Developer at Fangyuan Computer Software Co. Ltd.

Duties:

Helped design the layout for all kinds of written documents

Computer Skills

Languages: C, Visual C++, Java

Operating Systems: Windows XP/Windows 7/Windows 10

Software/Applications: Microsoft Office, Microsoft Access, Microsoft Visual C++, Microsoft Project, Microsoft Publisher, Novell Netware

Hardware: PC, HP-UX, Sun Sparc, Macintosh, SGI Crimson

Networking: Microsoft LAN Manager, TCP/IP, Novell Netware

Qualifications/Certifications: CET4/6, TOPE, NIT, MCSE, Undergraduate Diploma, Bachelor's Degree of Computer Science

Hobbies

Sports, Reading

A Covering Letter should accompany the CV with a completed application form (See Unit 2 Job Application Form). 求职信应该辅以简历。It helps to draw the employer's attention to the applicant's particular experience or skills which will be useful to the company. 求职信中求职者的具体工作经验和技能会吸引雇主的注意。If someone gets the employer's attention with a good covering letter, this will encourage him to read the applicant's CV in more detail. 一封好的求职信能够吸引雇主的注意并且能够使雇主仔细阅读求职者的简历。

A good covering letter should include the following points: 一封好的求职信应包括以下内容：

1. the job title you're applying for 应聘职位的名称
2. where you found out about it (which newspaper of their advertisement) 招聘信息来源
3. why you're interested in that type of work 求职原因
4. summarize your strengths and how they might be an advantage to the organization 求职者的优势及其对公司的好处
5. relate your skills to the job 有哪些与职位相关的技能
6. why the company attracts you 招聘公司为什么能够吸引你
7. when you're available to start work and for the job interview 提供可以参加面试的时间
8. say you look forward to hearing from them soon 表达要求收到回复的愿望

Sample 1: (without the inside address)

May 28th, 2020

Dear Sir/Madam,

With reference to your job advertisement in today's "Harbin Life Daily", I am interested in applying for the position of the network administrator with your company. Your advertisement addresses my qualifications perfectly. I can offer you the exact skills you are looking for in the applicants.

I studied Web Site Design & Development and Project Teamwork & Communication which are the knowledge that you require. I am also good at organizing project teams and distributing work tasks to team members appropriately with productive, individual contributions to the group.

I enclosed my CV.

Your company attracts me most is that you have a lot of potential. Also I prefer to work for a relatively small friendly company rather than a big cold corporation.

I am now preparing for my undergraduate diploma examinations and the degree dissertation oral defense. Therefore, I am able to start working in June, 2020.

Furthermore, I am available for the job interview at any time which is convenient for you.

Thanks for your careful consideration to my application. I look forward to hearing from you.

Yours faithfully,

Wang Jinsheng

Sample 2: (without the inside address)

August 28th, 2020

Dear Sir/Madam,

I am writing to apply for the position of the network administrator you advertised in today's "China Daily". My qualifications meet your requirements well. I have the proper skills you are looking for.

I studied Software Design, Database and Project Management which are the knowledge that you need. I am also able to organize project teams and allocate roles to team members properly to make productive working teams.

Please find the enclosed CV.

I am interested in your company because you have excellent reputation in this industry. Also I would like to work for your kind of professional company rather than a company with a negative image.

I am now preparing for myself fully for my future career. Therefore, I can start working any time from now on.

Moreover, I am available for the job interview at any time which is suitable for you.

Thank you for your careful consideration to my application. I look forward to hearing from you.

Yours faithfully,

Ma Shaolin

Exercise 6

Write your own CV and a covering letter.

CV

CV

Personal information (name, address, E-mail, contact numbers, etc. don't write too much)

Career target or job objective (list the positions you want to apply for to show your expectations)

Profile (brief your abilities, strengths, etc. use expressive adjectives + noun phrases)

Education (include the time, the degree or diploma, the major and the school, relevant coursework, etc.)

Work Experience (write the time, job title, the name of employer, duties/responsibilities, etc)

Other skills (computer skills and foreign language skills)

Qualifications/Certifications (list all degrees, diplomas, certificates)

Hobbies (mention the good ones which help enhance your image)

Unit 11 The Development Environment

Letter

June 20th, 2020

Dear Sir/Madam,

Further to your job advertisement in ………………………., I am writing to apply for the post of ……………………… with your company. The qualifications I have…....................................

I studied ……………………………………………………………………………. which you require. I am also good at …………………………………………………………….

Please find the enclosed CV.

Your company attracts me most is that ………………………………………….. Also I prefer to work for ………………………………………………………….. rather than ………………………………....

I am now ……………………………….……………………... Therefore, I am able to start working in …………, 2020.

Moreover, I am available for the job interview at any time which suits you better.

Thanks for ……………………………………………………………………………... I look forward to ……………………………….

Yours faithfully,
………………

Transcript

Listening 1
M: Creation software. What can I do for you?
L: Hi, I'm trying to use your Game Player Home Edition I bought the other day on my new XP system. It doesn't work. It always shuts down with a warning when it starts up!

145

M: What does the warning say then?

L: Wait a moment. Let me see. Oh, yeah, it prompts that a Java Virtual Machine should be installed.

M: Yes, that's it. Windows XP does not provide Java Running Environment, which is formerly provided in 98 and 2000. You must download a Java Virtual Machine and then install it in your system. You can visit the homepage of Sun Microsystems, and download the latest version of JRE.

L: And that will solve the problem?

M: Yes. You should have no further trouble.

L: Thank you so much.

M: Anytime. Have a nice day.

Listening 2

M: Good morning. Creation Software.

V: Hello. May I speak to Mr. George Black, head of the Technical Department, please?

M: I'm sorry. Mr. Black has just gone out. May I ask who is calling, please?

V: This is John Smith from Wall Consultant. Would you please tell me when I can get a hold of him?

M: I really have no idea when Mr. Black will be in the office. Could you please call back later, or would you mind leaving a message?

V: Would you mind telling him that due to some sudden changes, we want the MIS application you are developing to be programmed with C#.

M: OK. I will give him the message as soon as he is back.

V: Thank you very much.

Unit 12　New Technology

Section One　Reading

➢ Technical Reading

Have you ever heard of Machine Learning ? How does machine learning? Let's pass along the following article into the Machine Learning world.

What is **Machine Learning**

With the development of artificial intelligence technology, computers and electronic products are becoming more and more intelligent. Machine learning technology belongs to a branch of artificial intelligence, which is a kind of algorithm that can automatically analyze and obtain rules from data, and use the rules to predict unknown data.

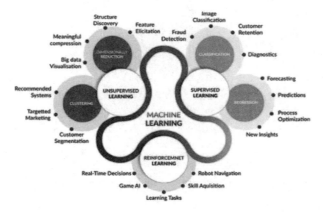

*Machine learning technology belongs to a branch of artificial intelligence. Its theory is mainly divided into the following three aspects: **supervised learning**, **unsupervied learning** and **reinforcement learning**.*

In our lifes, machine learning technology is mainly reflected in the following parts:
- Data Mining: Discover the relationship between data.

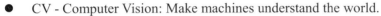

- CV - Computer Vision: Make machines understand the world.
- NLP - Natural Language Processing: Make machines understand the text.
- Speech Recognition: Make machines understand what we say
- Decision Making: Let machines make decisions, such as car control decisions in driverless vehicles.

Machine Learning algorithms are generally divided into two categories: Supervised Learning and Unsupervised Learning. Supervised Learning means that the machine learns a function from the given training data set, and when new data comes, it can predict the result according to this function. The training set of Supervised Learning is required to include input and output, also can be said to be features and objectives. The goal of the training set is marked by people. Common Supervised Learning algorithms include regression analysis and statistical classification. Unsupervised Learning is a kind of data processing method that the machine classifies the samples by analyzing the data of a large number of samples without class information.

In addition, there is an important branch of Machine Learning, Reinforcement Learning. The problem of Reinforcement Learning includes learning how to do and how to map the environment into action, so as to get the maximum reward. In Reinforcement Learning, learner is a decision-making agent, it will not be told what action to perform, but through repeated attempts to find the behavior that can get the maximum reward. In general, the action will affect not only the current reward, but also the environment at the next time point, so it will also affect all subsequent rewards.

Vocabulary

1. Machine Learning   n. 机器学习
2. Supervised Learning   n. 监督学习
3. Unsupervised Learning   n. 非监督学习
4. Reinforcement Learning   n. 强化学习

Exercise 1

True or False

1. (　) Supervised Learning means that the machine learns a function from the given training data set.
2. (　) Data Mining can make machine understand what we say.
3. (　) Reinforcement Learning is a kind of data processing method that the machine classifies the samples by analyzing the data of a large number of samples without class information.
4. (　) NLP can make machines understand the text.

Unit 12　New Technology

➢ **Fast Reading**

Text 1

iPhone Touch-screen

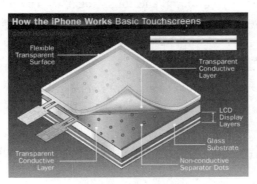

iPhone Touch-screen

　　Electronic devices can use lots of different methods to detect a person's input on a touch-screen. Most of them use sensors and circuitry to monitor changes in a particular state. Many, including the iPhone, monitor changes in electrical current. Others monitor changes in the reflection of waves. These can be sound waves or beams of near-infrared light. A few systems use transducers to measure changes in vibration caused when your finger hits the screen's surface or cameras to monitor changes in light and shadow.

　　The basic idea is pretty simple—when you place your finger or a stylus on the screen, it changes the state that the device is monitoring. In screens that rely on sound or light waves, your finger physically blocks or reflects some of the waves. Capacitive touch-screens use a layer of capacitive material to hold an electrical charge; touching the screen changes the amount of charge at a specific point of contact. In resistive screens, the pressure from your finger causes conductive and resistive layers of circuitry to touch each other, changing the circuits' resistance.

　　Most of the time, these systems are good at detecting the location of exactly one touch. If you try to touch the screen in several places at once, the results can be erratic. Some screens simply disregard all touches after the first one. Others can detect simultaneous touches, but their software can't calculate the location of each one accurately. There are several reasons for this, including:

- Many systems detect changes along an axis or in a specific direction instead of at each point on the screen.
- Some screens rely on system-wide averages to determine touch locations.
- Some systems take measurements by first establishing a baseline. When you touch the

screen, you create a new baseline. Adding another touch causes the system to take a measurement using the wrong baseline as a starting point.

The Apple iPhone is different—many of the elements of its multi-touch user interface require you to touch multiple points on the screen simultaneously. For example, you can zoom in to Web pages or pictures by placing your thumb and finger on the screen and spreading them apart. To zoom back out, you can pinch your thumb and finger together. The iPhone's touch screen is able to respond to both touch points and their movements simultaneously.

Exercise 2

Fill in the blanks according to the information from the text above.

1. Electronic devices can use lots of different methods to detect a person's input on a _____.
2. In screens that rely on _____, your finger physically blocks or reflects some of the waves.
3. In resistive screens, the _____ causes conductive and resistive layers of circuitry to touch each other, changing the circuits' resistance.
4. The _____ touch screen is able to respond to both touch points and their movements simultaneously.

Text 2

Big data is the term for a collection of data sets so large and complex that it becomes difficult to process using on-hand database management tools or traditional data processing applications. The challenges include capture, curation, storage, search, sharing, transfer, analysis, and visualization. The trend to larger data sets is due to the additional information derivable from analysis of a single large set of related data, as compared to separate smaller sets with the same total amount of data, allowing correlations to be found to "spot business trends, determine quality of research, prevent diseases, link legal citations, combat crime, and determine real-time roadway traffic conditions."

As of 2012, limits on the size of data sets that are feasible to process in a reasonable amount of time were on the order of exabytes of data. Scientists regularly encounter limitations due to large data sets in many areas, including meteorology, genomics, connectomics, complex physics simulations, and biological and environmental research. The limitations also affect Internet search, finance and business informatics. Data sets grow in size in part because they are increasingly being gathered by ubiquitous information-sensing mobile devices, aerial sensory technologies (remote sensing), software logs, cameras, microphones, radio-frequency identification readers, and wireless sensor networks. The world's technological per-capita capacity to store information has roughly doubled every 40 months since the 1980s; as of 2012, every day 2.5 exabytes (2.5×1018)

of data were created. The challenge for large enterprises is determining who should own big data initiatives that straddle the entire organization.

Big data is difficult to work with using most relational database management systems and desktop statistics and visualization packages, requiring instead "massively parallel software running on tens, hundreds, or even thousands of servers". What is considered "big data" varies depending on the capabilities of the organization managing the set, and on the capabilities of the applications that are traditionally used to process and analyze the data set in its domain. "For some organizations, facing hundreds of gigabytes of data for the first time may trigger a need to reconsider data management options. For others, it may take tens or hundreds of terabytes before data size becomes a significant consideration."

➢ Supplementary Reading

Text 1

Wi-Fi only or Wi-Fi + Cellular iPad?

There are two types of Apple iPad models: Wi-Fi only and Wi-Fi + Cellular. The Wi-Fi only model allows you to connect to Wi-Fi networks, while the other model supports both Wi-Fi and cellular data connections. Depending on the iPad model and service provider, cellular data access may be provided through either a 4G or 5G connection.

iPads that support both Wi-Fi and cellular data (sometimes called "cellular iPads") typically cost about $153 more than their Wi-Fi only counterparts. For example, a five-generation 128GB Wi-Fi only iPad is $456, while a 128GB cellular iPad is $609. Additionally, in order to access cellular data, you must purchase a monthly data plan from Verizon or AT&T. These data plans do not require a long-term contract, but they cost $15 to $50 per month.

If you have Wi-Fi access at home or work, the Wi-Fi only model is probably a good choice. As long as you have a Wi-Fi connection, you can download and upload data for free. Most ISPs offer unlimited data plans for home users, so you don't have to monitor your data usage either. For most home users, I recommend the Wi-Fi only model.

However, if you travel a lot and your iPad is your primary computer, the cellular model might make sense. While the Wi-Fi only iPad requires access to a Wi-Fi network, the cellular iPad allows you access the Internet from anywhere you have a cellular signal. If you are frequently on the go, the convenience of the cellular iPad may be worth the higher price and monthly fee.

Text 2

Android Overview

A software stack for mobile devices which Developed and managed by Open Handset Alliance. It is a Open-sourced under Apache License.

Android Features

- **Application framework** enabling reuse and replacement of components
- **Dalvik virtual machine** optimized for mobile devices
- **Integrated browser** based on the open source WebKit engine
- **Optimized graphics** powered by a custom 2D graphics library; 3D graphics based on the OpenGL ES 3.0 specification (hardware acceleration optional)
- **SQLite** for structured data storage
- **Media support** for common audio, video, and still image formats (MPEG4, H.264, MP3, AAC, AMR, JPG, PNG, GIF)
- **GSM Telephony** (hardware dependent)
- **Bluetooth, EDGE, 5G, and WiFi** (hardware dependent)
- **Camera, GPS, compass, and accelerometer** (hardware dependent)
- **Rich development environment** including a device emulator, tools for debugging, memory and performance profiling, and a plugin for the Eclipse IDE

Android ships with a set of core applications:

–E-mail Client

–SMS Program

–Calendar

–Maps

–Browser

–Contacts

- All applications are written using the Java language

Android Applications

- Android applications are written in Java
- The compiled Java code (along with any data and resource files required by the application) is bundled by the aapt tool into an Android package (.apk)

Each Android application lives in its own world:

–Runs in its own Linux process

- Started when any of the application code needs to be executed
- Shuts down when no longer needed or system resources are required by other applications

Unit 12 New Technology

  —Each process has its own virtual machine (sandboxing)
- Application code runs in isolation from all other applications

—Each application assigned a unique Linux user ID

One application can make use of elements of other applications

—Applications are composed of parts (components) that can be started up when the need arises, and instantiate the Java objects for that part

—Android applications do not have a single entry point (e.g. no main function)

—Essential components that the system can instantiate and run as needed

Four basic components: Activities, Services, Broadcast Receivers, Content Providers

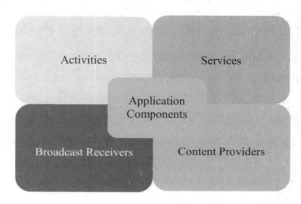

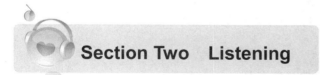

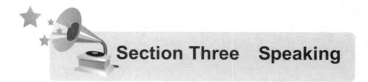

Exercise 3

What would you say in the following situations?

1. Imagine that you work at Sony company. You are answering a phone call. Play your role according to the clues given in the brackets. （假设你在索尼公司工作。现在你正在接听电话。根据括号中给出的提示完成角色扮演对话。）

  You: （电话铃响起，你接听电话，并说出 Sony 的客户服务部。）

153

Client: Hello, this is Sad Smith. I want to ask how to change the resolution of my T1 camera ?

You: （你可以通过相机屏幕旁边菜单键里面的选项进行调整，但是需要注意的是调整了分辨率之后照片的大小将会做相应的调整。建议你最好参照我们公司网站上的 FAQ 或者随机附带的用户手册进行操作。）

Client: Oh, I see. Thanks a lot!

You: （表示客气，并祝对方顺利。）

2. Imagine that you work at Apple company. You are answering a phone call. Play your role according to the clues given in the brackets. （假设你在苹果公司工作。现在你正在接听电话。根据括号中给出的提示完成角色扮演对话。）

You: （电话铃响起，你接听电话，并说出 Apple 的技术支持部。）

Client: Hello, this is Sun Smith. I want to ask how to update my music store.

You: （你需要到 www.apple.com 下载 iTunes 程序，安装到电脑上，用 USB 线将机器与电脑连接，电脑上会弹出 iTunes，此时 iPod 屏幕上会显示连接的状态。然后你可以点 iTunes 上面的选项，将歌曲、图片、电影等同步到 iPod 里面。到 http://www.apple.com/itunes/overview/ 上浏览一下 iTunes 的全部说明会是一个更好的选择。）

Client: Oh, let me try. Thanks a lot!

You: （表示客气，并祝对方顺利。）

Section Four Writing

Experience Certificate writing 工作经验证明书的写作

Sample 1:

Experience Certificate

To Whom It May Concern,

It is my pleasure to write on behalf of Mr. Wang Cheng who has worked with Hitech Computers in capacity of Java developer for the past 4 years.

During his time with us, Mr. Wang Cheng, has been a dedicated and valuable employee and he has worked hard at any and all tasks I have given him. He is quite confident and is a consummate professional. He has always exhibited sound judgment while in my employ and he is a trusted worker. He is quick to take initiative and I am very satisfied with his performance. He has been quite helpful in the advancement of our organization.

Furthermore, please allow me to expand on his conduct during his stay with us. He has a wonderful temperament and works very efficiently both independently or whilst working as part of a team. His co-workers are all pleased with him and they feel comfortable in teaming and coordinating with him to work toward common goals and objectives.

While he will be missed, I wish him all the best of luck in his future endeavors.

Yours faithfully,

…

Sample 2:

Experience Certificate

To Whom It May Concern,

It is my pleasure to write on behalf of Ms. Liu Jie who has worked with Acer Computers in capacity of Assistant Manager for six months.

During her work period Ms. Liu Jie remained involved in her work dedicatedly. I found her pretty active in whatever task I have provided her. She is a confident person. She is professionally sound, hard-working and devoted worker. She has the motivation to take initiative and I am gratified that she had been helpful in advancement of our organization.

Moreover, I would like to reflect over her conduct during her stay with us. She has a genial temperament and can efficiently work in a team. All my staff is pleased with her and feels comfortable in teaming and coordinating with her for the realization of organizational goals and objectives.

I wish her all the best in all her future career.

Yours sincerely,

Sample 3:

Experience Certificate

To whom it may concern,

This is to certify that Mr. Gao Wenjun was working at Online Science Corporation as a network maintenance clerk from Sep 2016 to July 2020.

During his tenure, we found him to be pleasant to work with. His services were found to be satisfactory.

We wish him all the best in his future endeavors.

Yours truly,

Exercise 4

Write a certificate of work experience.

Transcript

Listening 1

Kevin: Hello, this is Apple Customer Service. What can I do for you?

Customer: Hello, this is Tom speaking. I want to know how to transfer songs from my iPod to a Computer.

Kevin: When you connect your iPod to iTunes, you can enable the device as an external storage device under the Settings tab. Here's how you do that:
Connect your iPod to your computer and open iTunes.
Click on the iPod device in the iTunes source list on the left side of the screen.
Click the Settings tab at the top of the screen.
Under Settings, check the box that says "Enable disk use".
Close iTunes.
Locate the iPod icon in the Finder on a Mac or Explorer on a Windows format.
Drag the desired files over that icon.
Disconnect the iPod from the computer.
When you want to retrieve those files, perform the first four steps again.

Customer: OK, I got it. Thanks for the help.
Kevin: You're welcome. Thank you for using Apple products and services. Goodbye.

Listening 2

Kevin: Hello, Creative Company Support Center. Can I help you?

Customer: Hello, I bought your desk-computer last week. I want to know how to connect my computer to my TV.

Kevin: The biggest problem with connecting your computer to your TV is that, generally speaking, computers and TVs don't display at the same resolutions. Luckily, most HDTVs have the ability to scale incoming signals to match their native screen resolution. And two programs are considered the best solutions for solving connectivity problems between a computer and a TV: PowerStrip for Windows and DisplayConfigX for Mac. Both of these programs allow you to match your graphics card's resolution precisely with the native resolution of your TV.

Customer: Ah, yes. I got it. That will solve the problem. Thank you.

Kevin: You're welcome. For lots of more information about home theater systems, please contact to our hardware department. Thank you for using our products and services.

Grammar

Unit 1　Tenses 时态

一、英语 16 种时态的动词形式

	现在时	过去时	将来式	过去将来式
一般	do/does	did	will do	would do
进行	am/is/are doing	was/were doing	will be doing	would be doing
完成	have/has done	had done	will have done	would have done
完成进行	has/has been doing	had been doing	will have been doing	would have been doing

注意：母语是英语的人用"将来形式"来表示"将来时态"，"将来时态"已经不再被使用了。"将来形式"包括"时间表的将来""日程安排的将来""计划的将来"和"推测或一般将来"。

二、英语时态的用法

英语时态表达两层含义：时间和内容。

三、七种时态的主要用法

1. 一般现在时

一般现在时用来表达：

（1）一个习惯（经常发生的动作），如"I go to work on foot."。

（2）（时间表或日历中的）将来事件，如"The boss interviews all candidates this Friday morning."。

2. 现在进行时

现在进行时用来表达：

（1）说话时正在发生的动作，如"They are installing the latest-version operating system."。

（2）进行中的动作，如"He is learning Visual C++. NET these days."。

（3）计划好的日程安排，如"Leaders are checking our safety work tomorrow."。

3. 现在完成时

现在完成时用来表达：

（1）现在已经完成的动作，如"We have trained our staff to use the English-version operating system."。

（2）过去的持续了一段时间的动作，如"Larry has worked for IBM since he was 27 years old."（现在已经不在那儿干了）。

4. 现在完成进行时

现在完成进行时用来表达：

（1）过去发生且持续到现在的动作，如"Larry has been working for IBM since he was 27 years old."（现在还在IBM工作）。

（2）过去动作造成目前的结果，如"Sandy is disappointed because she has been checking the new software and found a lot of problems with it."。

5. 一般过去时

一般过去时用来表达过去发生而且已经完成的动作，如"The man was a technical support representative."。

6. 过去进行时

过去进行时用来表达过去某时正在进行的动作，如"The salesman was writing letters this time yesterday."。

7. 过去完成时

过去完成时用来表达过去一个时间点之前已经完成的动作，如"The general manager had finished his presentation before 2:00 pm yesterday."。

四、四种将来式的主要用法

1. will/shall：一般将来或预测的将来

一般将来或预测的将来用来表达：

（1）将来的事实或根据个人观点推测的可能发生的事情，如"He will come for lunch."。

（2）说话时立即做出的决定、打算或提议，如"I will study hard."。

We will employ two network team leaders.

They will do that for us.

2. going to：计划的将来

计划的将来用来表达：

（1）根据现实情况推测某事必然发生，如"It is going to break down without careful attention."。

（2）说话前就已经做好了的计划、决定或打算，如"The college is going to send teacher abroad to study computer science."。

3. future continuous 将来进行式

将来进行式用来表达将来某一时刻正在发生的动作，如"The meeting will be taking place at 10 am tomorrow."。

4. future perfect 将来完成式

将来完成式用来表达将来某一时间之前已经完成的动作，如"You will have finished your studies for the undergraduate course by May next year."。

注意：其他时态的用法参考所有现在各种时态的用法。

Exercise

Fill in the gaps with the proper tense form of the words given in the brackets.

1. At last, a lot of individuals have a website that _____ (provide) their personal information.
2. What achievements _____ you _____ (make) to your employer?
3. This job opportunity, if I am employed, _____ (give) me experiences on how to develop the overseas markets.
4. Since I _____ (be) a child, I _____ (develop) an interest in surfing the internet.
5. I _____ (study) in the UK this time next year.
6. I _____ (finish) my research work by the end of next year.
7. The World Wide Web _____ (be develop) in 1989 by Tim Berners-Less of the European Particle Physics Lab (CERN) in Switzerland.
8. I _____ (pass) CET4 when I graduated from the high school.
9. The machine will alarm when a false operation _____ (be conduct).
10. The plane _____ (leave) when I got to the airport. So I could barely bid him a farewell.

Unit 2　Passive Voice 被动语态

一、为什么用被动语态

当人们只关心宾语或者什么事情被做的时候，会使用被动语态，因为"是谁做的"已经不重要了。

二、被动语态的主要结构

被动语态的动词结构是：**be + p.p.**（p.p. 是动词的过去分词形式），而且 be 动词在具体情况下应该用恰当的时态。

三、各种时态被动语态的谓语动词形式

Be 动词的 16 种时态形式如下：

时间时态	现在时	过去时	将来式	过去将来式
一般	am/is/are + p.p.	was/were + p.p.	will be + p.p.	would be + p.p.
进行	am/is/are being + p.p.	was/were being + p.p.	will be being + p.p.（非常少见）	would be being + p.p.（非常少见）
完成	have/has been + p.p.	had been + p.p.	will have been + p.p.	would have been + p.p.
完成进行	have/has been being + p.p.	had been being + p.p.（非常少见）	will have been being + p.p.（非常少见）	would have been being + p.p.（非常少见）

Exercise

Fill in the gaps with the proper passive tense form of the words given in the brackets.

1. It is reported that a new robot _____ (design) by him in a few days.
2. Diamond _____ (discover) in Brazil in 1971.
3. Tim has been out of work since he _____ (dismiss) three years ago.
4. A large number of colleges and universities _____ (establish) since 1949.
5. The goods _____ (unload) when they arrived at the airport.
6. Most environment problems still exist because no effective measures _____ (take) in the past.
7. Pluto, the outermost planet of the solar system, _____ (discover) photographically in March 1930.
8. The construction of the laboratory _____ (complete) by the end of next year.

Unit 3   Sentences 句子

一、句子基本类型

1. 肯定句

   如：A computer is a tool.

   A computer does a lot of things for people.

   A computer can be connected with others.

2. 否定句

   如：A computer is **not** just a tool.

   A computer does **not** do all things for people.

   A computer can **not** work without an operating system.

3. 一般疑问句

如：**Is** a computer a tool?

Does a computer do a lot of things for people?

Can a computer be connected with others?

4. 特殊疑问句

如：**What** is a computer?

What does a computer do?

What can a computer do?

5. 反义疑问句

如：You like your computer teacher, **don't you**?

They won't break the mobile hard disks, **will they**?

I am late, **aren't I**? (conversational); **am I not**? (formal)

I don't think he can finish his work in time, **can he**?

6. 祈使句

如：Go to a computer shop.

Don't go to an Internet bar at night.

7. 感叹句

如：What a wonderful computer!

How beautiful the laptop is!

注意：特殊疑问词包括：what、who、when、where、why、how 等。

二、句子成分

句子成分包括主语、谓语（Verb 动词）、宾语、定语、状语、补语。

三、句子主要成分

句子的主要成分是主语、谓语（动词）。

四、从句或分句

从句包括主语和谓语，它是构成一个完整句子或句子一个部分的基础。

五、简单句和复杂句

简单句只包含一个主句。简单句的主要结构是：主语、谓语和宾语。我们经常用 S、V、O 来表示主语、谓语和宾语。

复杂句包含一个或多个主句以及一个或多个附属从句。复杂句包括名词性从句、状语从句、定语从句和其他类型的从句。

六、从句

1. 名词性从句

名词性从句包括主语从句、宾语从句、表语从句、同位语从句。

2. 状语从句

状语从句包括时间/地点/程度/目的/结果/原因/条件/让步/方式状语从句。

3. 定语从句

定语从句包括限制性定语从句和非限制性定语从句。

Exercise

I. Name the following sentences with a correct type: **Positive sentence; Negative sentence; Yes-No question; Wh-question; tag question; Imperative sentence or Exclamatory sentence:**

1. Computer games are interesting.
2. Laptops are not as expensive as they were in the past.
3. Is it an operating system?
4. What is the most popular computer brand?
5. The man is a computer programmer, isn't he?
6. Please take your lap top with you when you come.
7. What a wonderful make!

II. **What clause is contained in each of the following sentences?**

1. How they teach the students is completely a joke.
2. I think that computer-based teaching methodology will be widespread world wide.
3. It is true that students like chatting online more than talking to friends face to face.
4. The computer teachers who work in the Information Science Department are experienced in making ppts.
5. Harbin Normal University Star College, which situates in Harbin Hulan Limin Development zone, was founded in 2003.
6. Students like where their school is located.
7. Teachers get the most satisfaction when their students make great achievements.
8. If a computer is linked with the Internet, it will bring a lot of convenience to the user.
9. This computer performs as well as that one.
10. This computer is so old that no more functions can be added onto it.

Unit 4 Nominal Clauses 名词性从句

一、名词性从句种类

名词性从句包括主语从句、宾语从句、表语从句、同位语从句。

二、名词性从句结构

1. 主语从句

主句的主语是个简单句且由 that、if/whether 和其他特殊疑问词引导，这样的句子是主语从句。

如：**That** Professor Brown is attending the conference is an honour. (formal)

2. 宾语从句

主句的宾语是个简单句且由 that、if/whether 和其他特殊疑问词引导，这样的句子是宾语从句（that 引导词可以省略）。

如：You must ensure **that** data is stored correctly.

3. 表语从句

系动词后有简单句且由 that、if/whether 和其他特殊疑问词引导，这样的句子是表语从句（that 引导词可以省略）。

如：We are sorry **that** this unfortunate incident occurred.

4. 同位语从句

That 引导的从句接在抽象名词（或该名词的衍生词）后，用来说明该名词的内容。该结构是同位语从句。抽象名词有 belief(believe)、comment(comment)、confidence(confident)、discovery（discover）、doubt（doubt）、evidence（evident）、fact、fear（fear）、hope（hope）、indication（indicate）、idea、information（inform）、knowledge（know）、news、opinion、order（order）、problem、promise（promise）、proof（prove）、proposal（propose）、report（report）、rumor（rumor）、story（tell）、suggestion（suggest）、thought（think）、truth（true）、wish（wish）等。

如：We must face **the fact that** the Earth is steadily warming.

Exercise

Fill in the following gaps with appropriate linking words:

1. It is known _____ sitting in front of computers all day is really bad to our health.
2. _____ VIP stands for is Very Important Person.
3. He asked _____ they will come to repair our computers.
4. _____ one is suitable for us is still a question.
5. We agree to the opinion _____ we must reduce CO_2 release.

6. Please tell me _____ to enable the connection.
7. _____ he will start working is uncertain because of the preparation has been made.
8. _____ he will stay is not decided, it depends on how far away his customer lives.

Unit 5　Reported Speech 间接引语

一、直接引语和间接引语

当转述"他人所说的话"的时候，直接转述他人的原话叫直接引语，不直接转述他人的话叫间接引语。因此，需要学习如何把直接引语变成间接引语。

二、直接引语和间接引语的转换方法

具体的转换方法是：根据要转述的话的具体内容而定。例如转述陈述句、一般疑问句、特殊疑问句、祈使句时，直接引语变成间接引语的转换方法是不同的。

1. 直接引语中的**代词、时态、时间词语、地点词语、词语顺序**等，在间接引语中需要变化。
2. 转述一般疑问句时，需要使用 **if / whether**。
3. 转述特殊疑问句时，需要特殊疑问词作引导词。
4. 转述祈使句时，需要考虑使用 **must、should、ought to and let's** 等。

三、陈述句、一般/特殊疑问句和祈使句的直接引语变成间接引语不同情况举例

陈述句

类型	举例
direct speech 直接引语	"I speak Chinese."
reported speech 间接引语（时态不变）	He says that he speaks Chinese.
reported speech 间接引语（时态变化）	He said that he spoke Chinese.

问句

类型		举例
特殊疑问句	直接引语	"Why don't you speak Chinese?"
	间接引语	He asked me why I didn't speak Chinese.
一般疑问句	直接引语	"Do you speak Chinese?"
	间接引语	He asked me whether/if I spoke Chinese.

祈使句

类型	举例
直接引语	"Carol, speak Chinese."
间接引语	He told Carol to speak Chinese.

Exercise

Report the things Diane said, use reported speech.

1. "Polly works very hard."
2. "I am going to buy a laptop."
3. "Why don't they go with us?"
4. "Leave me alone."
5. "Is she a Star College student?"

Unit 6　Adverbial Clauses 状语从句

一、状语从句的定义

从句在句中作状语叫状语从句。状语从句用来表示时间、地点、程度、目的、结果、原因、让步、方式或条件。

二、状语从句的种类

状语从句	定义和用法	引导词	例句
时间	时间从句用来表示主句动作发生的时间	when, as soon as, before, after, by the time, by	**When** a problem occurs, we want to know about it. This provides complex input and display capabilities **when** needed, on another computer.
地点	地点状语从句表示主句动作发生的地点	where and wherever	I will never forget Seattle **where** I spent so many wonderful summers.
程度	程度状语从句表示主句动作与从句动作的程度的比较	as…as; so…as; than, etc.	He is not **so** intelligent **as** his brother. His wife is much younger **than** he is.
目的	目的状语从句用来说明主句动作的目的	that, in order that, so that, in case and lest	Recover all data to a point of know consistency **in case** the system fails. He worked overtime **in order that** he can catch up with his workload.
结果	结果状语从句用来表达主句动作所带来的结果	so that, so …that, such …that	The famine was **so** serious **that** thousands perished. She laughed in **such** a strange voice **that** everyone turned to look at her.

续表

状语从句	定义和用法	引导词	例句
原因	原因状语从句用来说明主句动作或行为的原因	because, as, due to; that the reason why	I am calling **because** I have a problem with my home desktop computer. Embedded systems are often mass-produced, **so** the cost saved may be multiplied by millions of items.
让步	让步状语从句表示条件退让一步，表示虽然，尽管，即使等概念	though, although, even though, while, whereas or even if	I am trying to use a Wireless LAN on my Windows XP system, **but I can't access the network.** **Since** the system is dedicated to a specific task, design engineers can optimize it and reduce the size of the product.
方式	方式状语从句表示动作的方式	as, as if, as though, in that, the way and how	You can do **as** you like. He looks **as if** he were scared. The old farmer nodded, **as though** he understood everything. Jim gave us a pleasant surprise **in that** he stood first in the competition.
条件	条件状语从句表示主句动作发生的条件	if, on condition that, unless	**If/When** I am home, my mother cooks dinner. She **will** fail the exam **unless** she studies hard. **If** we **studied** harder, we would learn it. He **would** have gone abroad **if** she had had enough money.

Exercise

Fill in the gaps with an appropriate conjunction:

1. They have lived in Harbin _____ they moved here in 2000.
2. _____ he is a pupil, he knows quite a lot about computers.
3. _____ she goes, she will be warmly welcomed.
4. _____ you cross over to the other side of the street, you must be careful.
5. Her mother loves me _____ I were her son.
6. This notebook is not as good _____ that one.
7. Here is my mobile number _____ you can phone me when you come here next time.
8. My boss was angry _____ we were late.
9. She didn't go to bed _____ she had finished her work.
10. The outer design of these two players is so alike _____ they can't tell the difference.

Unit 7 Relative/Attributive Clauses 定语从句

一、定语从句的意义和种类

定语从句用来详细说明说话者所指的人或物，分为限制性定语从句和非限制性定语从句两种。

限制性定语从句用来详细说明说话者所指的人或物，非限制性定语从句用来提供说话者所指的人或物的额外信息。

二、定语从句的结构

定语从句的结构分为三部分：先行词（被描述的人或事）、关系代词（代替先行词且在从句中充当句子成分的代词）和关系从句（描述先行词的句子）。

如：The term WWW is used to describe an interlinked system of documents **in which** a user may jump form one document to another in a nonlinear, associative way.

E-commerce has grown out of businesses **that** started to advertise their presence via a Website.（限制性定语从句）

The person **who** was involved in this incident has been put on probation and sent to a special training class.（限制性定语从句）

This computer **(which/that)** I bought from you can't connect to the Internet.（省略定语从句）

The dialect of SQL **(which/that** is) supported by Microsoft SQL Server is called Transact-SQL (T-SQL).（省略定语从句）

The Term "PC" is originally used to describe an IBM-compatible personal computer, **which has an Intel CPU as its microprocessor.**（非限制性定语从句）

Exercise

Fill in the gaps with an appropriate relative pronoun:

1. Electronic commerce is a system _____ includes not only selling and buying goods transactions but also transactions generating demand for goods and offering sales support.
2. The software _____ was written for embedded system is often called firmware.
3. The rules _____ are defining data relationships must not be violated.
4. What is the name of the computer department manager _____ software you borrow.
5. The computer centre _____ we went to was really beautiful.
6. Do you know the date _____ we have to hand in our assignment.
7. Do you know the reason _____ we have to study English grammar.
8. _____ wins will go on to play in the final.

Unit 8　Relative/Attributive Clauses and Appositive Clauses
定语从句和同位语从句

定语从句和同位语从句的区别

- 同位语从句通常在一个抽象名词后，这个抽象名词经常是动词或形容词的派生词。
- 定语从句用来确定和说明讲话者所专指的人或物。

试比较：

The story **that** I wrote was published.

The story **that** I had resigned was published.

第 1 句中的 that 可以用 which 替代，第 2 句则不可以。

第 1 句中的 that 指代 the story 而且是 wrote 的逻辑宾语。第 1 句还可以改写成两句："I wrote the story." "The story was published."

第 2 句中 I had resigned 是 the story 的内容。That 也不是 had resigned 的逻辑宾语。也就是不能说："I had resigned the story." 这句话也可以改写成两句："The story was published." "I had resigned."

Exercise

Identify the relative clause or appositive in each sentence.

1. Betty Hardy, a computer teacher, helped found the computer net bar at Star College.
2. My brother, who is a supervisor in his department, drives a car.
3. I took a cookie from Johnson, who is the principal's daughter.
4. I took a cookie from Johnson, the principal's daughter.
5. Alice, the head of Computer Science Department, was saved from the flood by the police.
6. I once saw Zhang Ziyi, the famous film star.
7. Maggie Green, who is a professional photographer, worked for a university in Beijing.
8. Lisa, a good country woman, has a daughter named Emily.

Unit 9　Modal Verbs 情态动词

一、情态动词的定义

情态动词用来表达说话者对主要行为动作的态度，如能力、责任或义务、建议、许可、可能性等。

二、情态动词的形式

尽管情态动词有现在和过去两种形式，但是情态动词没有时态的区分。情态动词的现

169

在形式指现在和将来情态动词的过去形式指过去、现在和将来。情态动词的完成形式也存在，也指过去、现在和将来，见下表。

肯定形式		否定形式	
will/would	do	will not/would not; won't/wouldn't	do
shall/should		shall not/should not; shan't/shouldn't	
may/might		may not/might not; mayn't/mightn't	
can/could		can not/could not; can't/couldn't	
must		must not; mustn't	
need		need not; needn't	
dare		dare not; daren't	
ought to		ought not to; oughtn't to	

注意：一般疑问句形式是情态动词提到主语前面（倒装）。

三、情态动词的用法

表能力的情态动词：can、could、be able to

如：I can use C++.

I am able to use a computer.

I could use C++ when I was ten.

表责任或义务的情态动词：(rules and laws) must/have to、mustn't、can't、should、shouldn't、couldn't

(personal authority) must/have to、mustn't、can't、will、won't、shall、shan't

如：

	obligation	lack of obligation
present	You must drive on the left in China. You must finish your work. You mustn't steal. You mustn't leaver earlier. You can't smoke in the computer centre. You can't stay up tonight. You should work 8 hours. You will finish your food before 12:00. You shouldn't rest for too long. You shan't go out again for a week.	You needn't pay tax if your salary is below 1,000 yuan. You needn't tidy your desk. You can go Into an Internet bar when you are over 18. You can work late if you like. You may sign your contract for a maximum of 5 years. You may have a bath every day.
past	Drivers couldn't drive without a license.	I thought you could drive without a license.

Grammar

表建议的情态动词：must (strongest)、should (strong)、ought to (less formal)、could (for general advice)

如：

	advice
present	You must switch off your computer after work. You should set up a firewall for your computer operating system. You ought to know about the latest software to keep up to date with computer science. You could learn it online.
past	The Department Head said you must/should/ought to program it with C++ instead of VB.

表许可的情态动词：may、can、might、could

如：

	permission
present	You may start your computer now. May I use your computer? You can go now. Can I use your keyboard? Could I borrow your ASP.NET book?
past	She said you could go. She said you might come in. She asked if she might interrupt the meeting.

表可能性的情态动词：will、must、should、may、might、could、can

从下表可以看出针对可能性的大小所选用的情态动词。

degree of probability	positive	negative
certain (a future fact)	will	will not/won't
certain because of a lot of evidence	must	can not/can't/could not/couldn't
reasonably expected to happen	should	should not/shouldn't
possible to happen	may	may not/mayn't
slightly less certain than "may"	might	might not/mightn't
slightly less certain than "might"	could	might not/mightn't
general possibility; logically possible ("Can" is not used to mean possibility in a particular situation)	can (not for future possibility)	can't

171

如：

	probability
present	I will be on a business trip next month. My secretary won't go with me. (Your computer is on so there must be someone using it.) It can't be me. He should be here soon. It shouldn't take long for us to see him. It may be your secretary. It may not be my secretary. It might be yours. It might not be my manager. It could be our boss. It might not be a person from outside our company. Leaving a computer on after work can be dangerous.
past	They will have installed a new computer for public use by lunch time tomorrow. The system must have been upgraded because it is quicker now than before. They should have done it earlier. We may have used it if they did it a month ago. You might have lost the documents in your computer. She could have found something strange to your computer.

Exercise

Fill in the gaps with an appropriate modal verb:

1. Sandy _____ drive but she doesn't have a car.
2. I used to _____ stand on my hands but I can't do it now.
3. My grandmother was an intelligent woman. She _____ speak 6 languages.
4. Passengers _____ go through the security check at the airport.
5. You _____ smoke in the non-smoking area.
6. I can stay in bed late because I _____ work tomorrow.
7. It is late. I _____ go now.
8. Driving a motorbike without wearing a helmet is dangerous. He _____ wear a helmet.
9. She is feeling sick. She ate too much. She _____ have eaten so much.
10. _____ I use your computer? Mine is not working.
11. Claire _____ get bored in her job. She does the same thing every day.
12. The boss should be here before 9:30. It's 10 now and he never comes late. He _____ be coming.

Unit 10 Inversion 倒装

一、主谓倒装

倒装包括助动词和主语的倒装。最常用的倒装形式是在疑问句中。

Grammar

Earth is small compared to other planets.
Is Earth small compared to other planets?
如果句子中没有助动词，在变成一般疑问句时，要加助动词而且改变次序。
She looked tired.
Did she look tired?

二、条件从句中的倒装形式

在正式英语条件句中，主语和助动词的倒装。

第二条件句

Were I you, I would visit grandfather.

Had I some money, I would buy myself a hotdog.

第三条件句

Had I known this fact before, I wouldn't have come here.

Had Jake been informed about the meeting, he would have participated.

三、倒装表示强调

Little did she know how much work has left.

On no account must you sleep at class.

Never should you forget who your boss is.

Only then can it belong to me.

Here comes the moon.

以上句子结构略显晦涩，这意味着书上可能出现过类似的语句。

四、so、neither 和 nor 等否定词的倒装形式

"I am not into classical music." "**Neither** am I." (Or: "**Nor** am I.")

"I am fond of watching films." "**So** am I."

五、否定副词后的倒装形式

David **rarely** speaks to himself.

Rarely does David speak to himself.

Seldom does Jack get invited to parties. (seldom = rarely)

Never have they seen such a fascinating view.

At no time did the prisoner look as if he might confess.

Not only is she a great singer **but** she is **also** an amazing chemist.

Not until she took up mountain climbing did she overcome her fear of heights.

Under no circumstances will prisoners be allowed to give interviews to the media.

Little did she realize that her grandfather was really a fox.

173

注意：如果副词不在句首，不用倒装。比较下面的句子。
She is not only a great singer but she is also an amazing chemist.

六、Hardly…when…; Scarcely…when…; No sooner…than…的倒装形式

下例句中描述一个动作发生后另一个动作立刻发生。

Hardly had he walked outside **when** it started to snow.

Scarcely had he walked outside **when** it started to snow.

No sooner had he walked outside **than** it started to snow.

Note that the past perfect tense is used to describe the event that happened first.

注意：过去完成时用来描述一个动作比另一个动作先发生。

七、Only 要求的倒装

下例句子中 only 要求的倒装结构。

注意：要求倒装的动词不总是第一个动词。

Only after he arrived at the railway station did he look for his train ticket.

Only if you look through this dark glass will you be able to see the spots on the sun.

We accepted the invitation. **Only later** did we suspect it might be a trap.

Only by threatening extreme physical violence was the teacher able to control the class.

注意：下面句子中的 only 不要求倒装：

Only Fiona knew the answer to the question.

八、As 倒装句

as 的倒装形式表示两个相同的事物。

Beth was too shy to dance, as was Peter.

She decided to leave early, as did Miller.

注意：在这些句子中，"as" 表示两个事物之间的相似性。

九、Such 和 so 的倒装句

So excited were they that they couldn't sit still.

Such was their excitement that they began to jump up and down.

注意：so 后接形容词，such 后接名词。

十、形容词后的倒装形式

一些非常文学化的句子通常以形容词开头，包括倒装句。例如：

Blessed are the children who are still unaware of what the future holds.

Gone are the days when I could have been happy.

注意：这里过去分词的用法和形容词一样。

Grammar

十一、May 的倒装形式

当我们许愿时，我们可以用倒装法。例如：
May you both live happily ever after!

十二、感叹句的倒装形式

Aren't you a silly girl!
Isn't it a lovely day!

十三、副词倒装形式

Into the room ran the lady.
First comes love, then comes marriage.
After A comes B, then comes C, next comes D.
Down came the rain and washed the spider out.

十四、The more…, the more…倒装句型

The closer an object is to another object, the greater is the gravity between the two objects.

十五、"Story speech" 故事讲述时用的倒装形式

"I think it's time to go, " said Susan.
"It's time for you, but not for me, " replied Gary.
"Maybe we should collect our thoughts for a moment, " commented Ian.

注意：这种情况属于可选择的一种倒装形式，既可以倒装，也可以选择不用倒装。

Exercise

Fill in the gaps with an appropriate word (See the above example sentences):

1. _____ I you, I would study hard.
2. _____ I studied hard in the past, I would have gone to a better university.
3. _____ did she know about the UK, she made a lot of stupid mistakes.
4. _____ should you forget about your mother's birthday.
5. Only _____ can we go home.
6. _____ comes the teacher.
7. _____ does the tutor come to class.
8. _____ only is she a teacher but she is also a boss.
9. Hardly had he arrived at school _____ the bell rang.
10. Belinda was too quiet, _____ was Dan.
11. _____ are the days when I could have been happy.

175

12. _____ you go to pick a plum for my daughter.
13. The harder you study, the _____ results you will get.
14. _____ you an honest girl!
15. _____ you both succeed in the NIT exams!

Unit 11　Non-finite Verbs 非谓语动词

一、非谓语动词的种类

非谓语动词有三种：动词不定式、分词和动名词。

二、动词不定式

1. 动词不定式的形式

一般形式：to+动词原形。

时态	语态	
	主动语态	被动语态
一般式	to do	to be done
进行式	to be doing	—
完成式	to have done	to have been done

2. 动词不定式的用法

（1）作主语。

　　To learn computer science needs our great effort.

（2）作表语。

　　The function of the modem is to link more computers.

（3）作宾语。

　　Linus decided to develop a system that exceeded the Minix standards.

（4）作宾语补足语。

　　Page printers use laser to achieve high-speed hard-copy output.

（5）作定语（动词不定式作定语时放在被修饰词的后面）。

　　He has no chance to pass the exam.

（6）作状语。

　　To build your own computer, you would need these parts.

3. 动词不定式的否定结构

动词不定式的否定结构：not +to+动词原形。

Linus decided not to develop a system that would exceed the Minix standards.

4. 动词不定式的时态

（1）She seems to know about the problem.（一般式表示不定式动作发生在谓语动作之后）

（2）The staff pretended to be working.（进行式表示不定式动作与谓语动作同时发生）

（3）The client said to have arrived here by 8:30.（完成式表示不定式动作发生在谓语动作之前）

5. 动词不定式的语态

（1）He needs to repair the laser printer.（主动）

（2）The laser printer needs to be repaired.（被动）

6. 特殊情况

（1）The employee has nothing to do but play.（but 前面有 do 时 but 后省略 to）

（2）You'd better get the permission before you leave. （had better 后省略 to，还有 had best、would rather、would sooner…than…、can't but 等后面的 to 省略）

三、分词

1. 分词的种类、形式和特征

分词	动词 时态	语态	及物动词		不及物动词
			主动语态	被动语态	主动语态
现在分词	一般式		doing	being done	doing
	完成式		having done	having been done	having done
过去分词	一般式		—	made	done

2. 分词的用法

（1）作表语。

　　The lap-top computer is working.

　　He is interested in computer games.

（2）作定语。

　　The clerk caught a falling cordless keyboard.

　　The clerk found the lost cordless keyboard.

（3）作宾语补足语。

　　He heard the computer making a noise.

　　They noticed their notebooks touched by some dirty hands.

（4）作状语。

　　Entering the office, the boss was shocked.

　　Shocked by the shout, he didn't enter the office.

3. 分词的否定结构

分词的否定结构：not+分词。

（1）Not knowing what to say, he kept silent.

（2）Not told what to do, he did nothing.

4. 分词的时态

（1）He came to work, bringing his own notebook.（同时）

（2）Having finished his work at home, he came earlier than others.（分词动作在前）

5. 分词的语态

（1）The customer being invited is a VIP.（现在分词被动形式表示动作正在进行）

（2）Having been rebuilt, the computer started working again.（现在分词被动形式表示动作已经完成）

6. 特殊情况

gone、come、fallen、risen、arrived、grown、returned、passed、changed 等作表语时带有书面语色彩，如：

（1）Summer is gone.

（2）The leaves are fallen.

四、动名词

1. 动名词的形式

一般形式：动词+ing。

时态	语态	
	主动语态	被动语态
一般式	doing	being done
完成式	having done	having been done

2. 动名词的用法

（1）作主语。

　　Playing computer games is great fun.

（2）作表语。

　　Your job is designing computer games.

（3）作宾语。

　　Boys enjoy playing Internet games.

（4）作定语。

　　Make some dancing actions for the virtual model players.

3. 动名词的否定结构

动名词的否定结构：not+动名词。

The girls regret not learning to play computer games.

4. 动名词的时态

（1）We like chatting online.（同时）

（2）I forgot having met them before.（分词动作在前）

5. 动名词的语态

（1）The user doesn't like being asked to key in pass words for too many times.（动名词被动语态一般式表示动作与谓语动作同时发生或在谓语动作之前发生）

（2）The customer is delighted to have been given a prize after he purchases that computer.（动名词的被动语态完成式表示动作在谓语动作之前发生）

6. 特殊情况

在多数情况下都避免使用动名词被动语态完成式，而是用一般式代替，以免句子显得累赘，尤其是在口语中。

I remembered once being driven (having been driven) from Harbin to Changchun.

Exercise

Fill in the gaps with an appropriate form of the verb given:

1. I am a computer seller. I go home when I have finished _____ my computers. (sell)
2. After _____ all the doors, he went home. (check)
3. We saw a keyboard _____ from the shelf. (fall)
4. _____, she saw a stranger passing by immediately. (surprise)
5. Have you got anything _____ before you leave. (say)
6. You'd better _____ than not do. (do)

Unit 12　Infinitives&Gerunds 不定式&动名词

一、动词后面只接动词不定式的搭配

只接动词不定式的动词有 agree、decide、expect、hesitate、hope、want、plan、manage、wish、offer、prepare、refuse、fail、intend、pretend、neglect、propose、learn、promise、attempt、afford、would like 等。

如：The parents decide to buy a notebook for their child.

The student can't afford to buy himself a personal computer.

I would like to invite you out for lunch.

二、动词后只接动名词的搭配

只接动名词的动词有 like、feel like、dislike、fancy、enjoy、admit、mind、practice、suggest、regret、deny、avoid、imagine、miss、consider、delay、finish、resent、risk、appreciate、can't help 等。

如：The customers feel like using this kind of USB disks.

I suggest having a visual conference.

They can't help laughing when hearing him snoring.

三、动词后既接不定式又接动名词（表达的意义区别不大或者没有区别）

这类动词有 begin、start、intend、continue、bother 等。

如：It starts raining/It starts to rain.

Joe intends buying a printer/Joe intends to buy a printer.

Don't bother switching on the loudspeaker/Don't bother to switch on the loudspeaker.

四、动词后既接不定式又接动名词（表达的意义区别很大）

这类动词有 forget、remember、try、stop、regret 等。

如：He forgot taking his USB disk away.（忘记了做过……）

He forgot to take his USB disk away.（忘记了该去做……）

He remembers saving his document.（记得做过……）

Remember to save your document.（记得该去做……）

I tried taking the exam but failed.（试着做过……）

I tried to take the exam.（尽力去做……）

五、Do 和 doing 的区别

do 表示动作全过程，doing 表示动作的片段。

这类动词有 see、watch、notice、hear、feel、notice、smell 等。

如：We saw him playing basketball.

We saw him play basketball.

Tom heard the child shouting for help.

Tom heard the child shout for help.

We could smell the cake baking in the kitchen.

We could smell the cake bake in the kitchen.

Exercise

Fill in the gaps with an appropriate form of the verb given:

1. He likes _____. (swim)
2. They would like _____. (swim)
3. We felt someone _____ in the corridor. (cry)
4. They admit _____ him before. (meet)

Appendix A Glossary of Readings

英文词汇 | 中文解释

A

abbreviate | *vt.* 缩略
access | *n.&vt.* 存取
accounting | *n.* 账目，会计业务
adapter | *n.* 适配器
administrator | *n.* 管理者
all-in-one multifunction peripheral | *n.* 多功能一体机
alphanumeric | *adj.* 文字数字的
amazingly | *adv.* 令人惊讶地
analyze | *vt.* 分析，分解，解释
antiviral program | *n.* 反病毒程序
application | *n.* 应用程序
application software | *n.* 应用软件
argument | *n.* 范围
arithmetic and Logic Unit(ALU) | *n.* 算术逻辑单元
arithmetic operators | *n.* 算术运算符
ASCII | *n.* 美国信息交换标准代码
assembly language | *n.* 汇编语言
assume | *vt.* 假设，臆断，假装
audience | *n.* 观众，听众，读者
audio | *n.* 声音剪辑

B

backbone | *n.* 骨干，支柱
backup | *n.* 备份
binary | *n.* 二进制
bit | *n.* 字
blossomed into | 发展成，长成
byte | *n.* 字节

英文词汇	中文解释
C	
cascading style sheet (css).	层叠式样式表
catalog	n. 目录，目录册
category	n. 分类
CD-RW (compact disc- rewritable)	n. 可重写光盘
cell	n. 单元格
cell address	n. 单元格地址
Central Processing Unit (CPU)	n. 中央处理器
challenge	vt.&n. 挑战
character	n. 字符
chart	n. 图表
chip	n. 芯片
clip art	n 剪贴画
collaboration	n. 合作，协作
command	n. 命令，指令
communication	n. 通信
Communications software	n. 通信软件
community	n. 社区，社会，团体，共有，共享
compatible	adj. 兼容的
compile	n. 编辑
compiler	n. 编译程序
conduct	vt. 从事
confidential	adj. 秘密的，机密的
connection	n. 连接，联结，联系，关系，连接点
consistency	n. 一致性，稳定性
content template	n. 内容模板
context	n. 上下文，语境
correct	adj. 正确的
co-worker	n. 共同工作者，合作者，同事，帮手
crash	vt.&vi. 使猛撞，使撞毁
cursor	n. 光标
customize	vt. 定制，定做
cut	vt.&n. 剪切
D	
data type	数据类型
database	n.数据库，资料库

英文词汇	中文解释
DBMS (database management system)	n. 数据库管理系统
decryption system	n. 解密系统
delete	vt.&vi. 删除
dependent	adj. 依赖于
design template	n. 设计模板
designate	vt. 标明，指明
desktop	n. 桌面
dialect	n. 方言，语调
Digitizer table and pen	n. 图形输入板
disk	n. 磁盘，光盘
diskette	n. 软盘
display	vt. 列，显示，展览
document	n. 公文，文件，文献
dot-matrix printer	n. 点阵式打印机
download	v. 下载
duplicate	vt.&n. 复制（品）

E

element	n. 元素
encoded	vt. 编码
encryption	n. 密码
entire	adj. 全部的，整体的
erase	vt. 清除
established	adj. 确定的
execute	vt. 执行

F

facilitate	vt. 使容易，使便利
feature	n. 特征，特点，特色
field	n. 字段
field size	n. 字段宽度
file name	n. 字段名
financial statement	n. 财务报告
firewall	n. 防火墙
FireWire port	n. 火线接口
flash drive	n. 闪存盘
floppy	n. 软盘

英文词汇	中文解释
font specification	n. 字体规范
format	vt. 使格式化，编排格式
form	n. 窗体
formula	n. 准则，原则，公式
forward	v. 传输，传送
FTP (File Transfer Protocol)	n. 文件传输协议
full-motion video	n. 视频录像
function	n. 函数

G

global	adj. 全球的
graphic	n. 图形
Graphics software	n. 图形软件
GUI (graphical user interface)	n. 图形用户界面

H

handheld	adj. 手持的
hard disk	n. 硬盘
hardcopy	n. 硬拷贝
high-level language	n. 高级语言
http (Hypertext Transfer Protocol)	n. 超文本传输协议
hypertext	n. 超文本

I

illustrate	vt. 给……加插图，说明，阐明，表明
image	n. 图像
incorporate	vt. 包含，把……合并
individual	n. 个人
initial	adj. 最初的，开头的
innovative	adj. 创新的，革新的
input device	n. 输入设备
instant messaging	n. 即时消息，即时通信
instruction	n. 指令
Internet Explorer	微软公司出品的 Web 浏览器
intranet	n. 内联网
intuitive	adj. 直观的

J

joy-stick	n. 操纵杆

英文词汇	中文解释
K	
kernel	n. 内核，核心
keyboard	n. 键盘
L	
label	n. 文本，标签
LAN(Local Area Network)	n. 局域网
launch	vt. 启动
LCS(liquid crystal shutter)	n. 液晶体
LED(light-emitting diode)	n. 发光二极管
leg up	一臂之力
Linux	n. 一种开源操作系统
low-level language	n. 低级语言
M	
Mac OS	n. 苹果操作系统
machine code	n. 机器代码
Macintosh	n. 麦金塔电脑，Apple 公司于 1984 年推出的一种微机
malicious	adj. 恶意的
margins	n. 页边距
memorize	vt. 记住，熟记
memory	n. 内存
Microsoft Windows family	n. 微软家庭版视窗操作系统
Mobile	adj. 移动的
modem	n. 调制解调器
monitor	n. 显示器
multimedia projector	n. 多媒体投影仪
N	
nested	adj. 嵌套的
Netscape	美国 Netscape 公司，以开发 Internet 浏览器闻名
network	n. 网络
network card	n. 网卡
newsletter	n. 通讯，简报
non-text	n. 非文本

英文词汇	中文解释
note	n. 备注
numeric	adj. 数字的

O

object program	n. 目标程序
operating system	n. 操作系统
outline	n. 大纲
output device	n. 输出设备

P

Packet filtering	n. 数据包过滤器
Page printer	n. 激光页码式打印机
paragraph	n. 段落
parenthesis	n. 圆括号
paste	vt. 粘贴
perform	vt.&vi. 执行，表演，扮演
peripheral	n. 外围设备
Personal information management and personal finance software	n. 个人信息管理和财务软件
phenomenon	n. 现象
poses	vt. 造成，引起
precaution	n. 预防措施
presentation	n. 提供，显示，外观，报告，表演
Presentation software	n. 演示文稿软件
primarily	adv. 主要地，首要地
printer	n. 打印机
privacy	n. 隐私，秘密
private	n. 个人，私人 adj. 私人的，个人的
processor	n. 处理机
projected onto	n. 把……投影到
prompt	n. 提示，提醒
protocol	n. 协议
provide	vt.& vi. 提供，供给，供应
Proxy service	n. 代理服务器服务
public	n. 公众，大众 adj. 公众的，大众的
publish	vt.&vi. 出版，公布

英文词汇	中文解释
R	
Random-Access Memory(RAM)	n. 随机存取存储器
range	n. 范围
record	n. 记录
refund/exchange period	n. 退款/换货期
regulations	n. 规定，法规
relational database	关系型数据库
release	vt. 释放，开放；发布，发行
resources	n. 资源
respond	vt. 反应
retrieval	n. 检索
retrieve	vt. 恢复
reveal	vt. 显露，泄露
S	
scanner	n. 扫描仪
screen	n. 显示器
security	n. 安全
slide	n. 幻灯片
SMTP	abbr. 简单邮件传输协议
software	n. 软件
SOHO	家庭办公室，小型家庭公司
sort	vt.& vi. 分类
source program	n. 源程序
spreadsheet	n. 电子数据表
Spreadsheet software	n. 电子表格软件
Stateful inspection	n. 动态数据包过滤
stimulate	vt. 激发，促进
storage	n. 存储，存储器
synchronize	vt. （使）同步
syntax	n. 句法
System software	n. 系统软件
T	
take advantage of	v. 利用
TCP/IP（Transmission Control Protocol/Internet Protocol）	传输控制协议/因特网互联协议

英文词汇	中文解释
template	n. 模板
temporary	adj. 临时的，暂时的
Trojan horse	n. 特洛伊木马
typewriter	n. 打字机

U

update	vt. 更新
URL（Uniform Resource Locator）	n. 统一资源定位器，网址
USB（Universal Serial Bus）port	n. 通用串行接口
user-friendly interface	n. 用户友好界面
Utility software	n. 工具软件，通用软件

V

virus	n. 病毒
vulnerable	adj. 易受到攻击的

W

Web authoring software	n. 网络编辑软件
web browser	n. 网页浏览器
Web sites	n. 网站
wireless	adj. 无线的
Word processing and desktop publishing software	n. 文字处理和桌面出版软件
word wrap	n. 自动换行
worms	n. 蠕虫

Appendix B Key to Exercises

Unit 1 First Day at Work

Exercise 1

1. F   2. F   3. T   4. T   5. T

Exercise 2

1. b   2. a   3. e   4. d   5. c

Exercise 3

1. Dell Inspiron 660s   2. HP Z420 Workstation
3. 27-inch iMac   4. Dell Inspiron 15

Exercise 4

1. Inspiron Mini 9   light weight   plenty of battery life
2. Inspiron 13   processor speed   storage capacity
3. Inspiron 13
4. Inspiron Mini 9

Exercise 5

1. b   2. a   3. e   4. c   5. d

Exercise 6

A: Allow me to <u>introduce myself</u>. My name is Wu Min, a manager in the Sales Department.
B: <u>How do you do</u>, Miss Wu? Please to meet you.
A: <u>Nice to meet you</u>, here's my card.
B: Thank you. This is mine.
A: Thank you. <u>Let me introduce my colleagues here</u>. This is Miss Li, my secretary.
B: Glad to meet you.
A: This is Mr. Chen.

C: Nice to meet you.

Exercise 7

A: Wang Xin, <u>you haven't met</u> Mr. King, head of the England delegation, have you?
B: No, not yet.
A: Well, <u>come over and I'll introduce you</u>. Hello, Mr. King, I hope you're enjoying the party.
C: Yes, very much.
A: Mr. King, <u>I'd like you to meet a colleague of mine</u>. Mr. Wang, from our R&D Department.
C: <u>How do you do</u>?
B: How do you do? I'm very glad to meet you, Mr. King.

Exercise 8

Answer the questions by yourself.

Exercise 9

1. Memo

MEMO

To: All representatives
From: Miss Xinyi Dong, Office Manager
Subject: Ordering of company-headed stationery
Date: April 15th, 2009

The supply of company-headed writing paper, notepads and ballpens with the company's name and address will be available from next Monday from my office. These are to be distributed to customers.

All representatives should decide how many of these items they require. I need to have all orders on my desk in writing by this Friday at the latest.

---

2. Note

Danny,

Thanks for agreeing to do the interview for me next Tuesday. Here's the candidate's CV – I've marked a few things you should ask her about. And don't forget to take notes!

Thanks.

Amanda Ribera

---

3. Notice

> **Notice**
>
> Our electricity supply will be turned off tomorrow between 9am and 3pm.
>
> Staff may use batteries-powered machines during this period to continue prioritized work or arrange alternative tasks.
>
> I apologize for any inconvenience this may cause.
>
> Jay Yang, General Affairs Manager
> September 9th, 2008

Unit 2 Software

Exercise 1

1. a   2. c   3. h   4. i   5. g   6. j   7. b   8. f   9. e   10. d

Exercise 2

1. Mirocsoft's disk operating system.   2. COPY.
3. IBM.   4. Bell Laboratories.

Exercise 3

1. D   2. C   3. C   4. B   5. C

Exercise 4

1. D   2. C   3. E   4. G   5. B   6. H   7. F   8. A

Exercise 5

Answer the questions by yourself.

Exercise 6

Sample answer:

CITY BANK ACCOUNT APPLICATION FORM			
Surname	Wang		
First name	Qinglin		
Gender	Male		
Date of birth	October 15th, 1982		
Country of Origin	China		
Present address	66 Hongxiang St, Daoli District, Harbin		
Post code	150465		
When did you move to this address?	August 2014		
Permanent address (if different from above)	—		
Telephone no.(home)	0451-869186××		
Telephone no.(office)	0451-446052××		
Marital Status	Single		
No. of dependent children	0		
Residential details (owned or rented)	rented		
Employment status	full-time/part-time/self-employed/unemployed/retired		
Income details (per year)	￥100,000		
Signature	Qinglin Wang	Date	August 10th, 2014

Unit 3　Office Routine

Exercise 1

1. Graphical user interface.
2. GUIs make the computer much more user friendly and more suited to the casual IT user.
3. Windows, Icons, Mice and Pointers.

Exercise 2

1. F　2. F　3. T　4. F　5. T

Exercise 3

1. T 2. F 3. T 4. F 5. F

Exercise 4

A.
1. <u>WYSIWYG</u> stands for 'what you see is what you get'. It means that your printout will precisely match what you see on the screen.
2. <u>Justification</u> refers to the process by which the space between the words in a line is divided evenly to make the text flush with both left and right margins.
3. You can change font by selecting the font name and point size form the <u>type style</u>.
4. <u>Font menu</u> refers to a distinguishing visual characteristic of a typeface; "italic", for example is a <u>type style</u> that may be used with a number of typefaces.
5. The <u>format</u> menu of a word processor allows you to set margins, page numbers, spaces between columns and paragraph justifications.
6. <u>Mail merging</u> enables you to combine two files, one containing names and addresses and the other containing a standard letter.
7. An <u>indent</u> is the distance between the beginning of a line and the left margin, or the end of a line and the right margin. Indented text is usually narrower than text without <u>indent</u>.

B.
1. b 2. f 3. a 4. d 5. c 6. e

Exercise 5

1. secretary
2. discuss, counseling program
3. President Li
4. the sales manager
5. transfer

Exercise 6

1. Peter
2. Good morning
3. I'm fine
4. and you
5. Absolutely
6. Why do you ask
7. That's great

8. I have another engagement to attend tonight

Exercise 7

Sample answer:

	1 <u>Writer's address</u>
	2 <u>Date</u>
3 <u>Receiver's address</u>	
4 <u>Salutation</u>	
5 <u>Body</u>	
6 <u>Complimentary close</u>	
7 <u>Signature</u>	

Unit 4   Creative Software

Exercise 1

1. A particularly important feature of desktop publishing systems is called   <u>WYSIWYGs.</u>
2. Service bureau offer services such as   <u>   offset printing   </u>.
3. PostScript fonts were created by   <u>   Desktop publishing   </u>.
4. Fonts refers to the style and size of a   <u>   typefaces   </u>.

Exercise 2

1. Computer graphics are pictures and drawings produced by computer.
2. CAD: Computer Aided Design
   CAE: Computer Aided Education
   CAM: Computer Aided Manage
3. In the car industry, CAD software is used to develop, model and test car designs before the actual parts are made, this can save a lot of time and money.
4. People in business can present information visually to clients in graphs and diagrams.

Exercise 3

1. b   2. c   3. e   4. d   5. a

Exercise 4

Answer the questions by yourself.

Exercise 5

Sample answer

Resource Planning Manager: Suitability for Home-based Working

Introduction
The purpose of this report is to assess the suitability of my position as Resource Planning Manager for home-based working.

Findings
My working pattern and that of my colleagues varies from week to week. During certain periods a large proportion of my time is spent doing fieldwork. This is followed by office-based work collating and recording the data collected. Once the results have been recorded, I proof-read the colour copies of all reports and maps.

As regards communication with workmates, department meetings are held twice a month. At all other times, face-to-face individual communication can be handled by phone. Most of the official meetings can be achieved whether I am in the office or working elsewhere.

Conclusion
It is clear that I would be able to undertake the same duties while working from home. Clearly, some days would need to be spent in the office. It would also be necessary to use the technical facilities of the office such as a networked computer and a printer at times.

Recommendations
I would suggest that I should be given the necessary equipment to work partially from home for a trial period. After this time, further consultation should take place in order to reassess the situation.

Unit 5 Communicate Online

Exercise 1

1. T   2. F   3. T   4. T

Exercise 2

1. Yes, they can.
2. If you're writing a message and want to finish it later, tap Cancel, then tap Save Draft.
3. Add important people to your VIP list, and their messages all appear in the VIP mailbox.
4. Favorite mailboxes appear at the top of the Mailboxes list. To add a favorite, view the Mailboxes list and tap Edit.

Exercise 3

1. C   2. C   3. A

Exercise 4

1. Creative Software, how to access the Internet
2. not wrong
3. general
4. advise the customer to refer to Users' Manual

Exercise 5

Answer the questions by yourself.

Exercise 6

Sample answer:

From: Li Mingliang@126.com	
To: customerzheng@freemail.com	
CC:	
BCC:	
Subject: What color do you like better?	
Date:	
Dear Customer,	

Thank you for the order you placed for our latest range of computer notebooks. I would be delighted if you could tell about the colors of your preferences. Then we will prepare the delivery for you.

If you have any more questions concerning our products, please do not hesitate to contact us.

I will wait for your confirmation of the color.

Yours truly
Li Mingliang
Sales Representative

Unit 6　Surf the Network

Exercise 1

1. e　　2. f　　3. c　　4. b　　5. a　　6. d

Exercise 2

1. new wireless PC Cards
2. the best connection available
3. information, equipment, staff
4. her boss, contacts

Exercise 3

1. The dialogue most probably take place in a sales department (or in a show).
2. One product is introduced.
3. It is three times faster than the others.
4. No.

Exercise 4

Answer the questions by yourself.

Exercise 5

Sample answer:

From: talentc@yahoo.com.cn
To: customer@gmail.com
CC:
BCC:
Subject: Invitation to our company end-of-year dinner party
Date: Dec 22nd, 2009

Dear Mr. Zhou

I would like to invite you to join our company's end-of-year dinner party.

We will celebrate the New Year at our dinning hall at 6:00 pm on December 30th, 2008. Please dress casually. It is a general corporate event. Please just bring yourself, no gifts.

We look forward to seeing you then.

Yours sincerely

Xiaoran Han
General Manager

Accepting

From: customer@gmail.com
To: talentc@yahoo.com.cn
CC:
BCC:
Subject: acceptance of invitation
Date: Dec 22nd, 2009

Dear Ms. Han

I was delighted to take part in your company's end-of-year dinner party.

As an exchange, I will definitely invite you to come to visit my company in the coming year. Please keep this in mind.

I look forward to your party.

Yours sincerely

Jia Zhou
General Manager

Appendix B Key to Exercises

Declining

From: customer@gmail.com
To: talentc@yahoo.com.cn
CC:
BCC:
Subject: reply to your invitation
Date: Dec 22nd, 2009
Dear Ms. Han It was very kind of you to invite me to your company's end-of-year dinner party. However, I am afraid that I am unable to make it because of my business trip to Europe during that time. Instead, I wish you every happiness and success in the new year. Yours sincerely Jia Zhou General Manager

Unit 7 Selling Products

Exercise 1

1. A database is used to <u>access, retrieval, and use of data</u>.
2. Information is entered on a database via <u>fields</u>.
3. A <u>relational database</u> has links between its files.
4. An <u>entity</u> is a subject about which information is stored in a table.
5. The <u>Database Administrator(DBA)</u> manages the database.

Exercise 2

1. Student, course.
2. Course id.
3. Course id.

199

Exercise 3

1. By <u>Back-up and recovery</u>, information in the database must not be lost in the event of system failure.
2. The DBMS must check <u>passwords</u> and allow appropriate <u>privileges</u>.

Exercise 4

Answer the questions by yourself.

Exercise 5

Answer the questions by yourself.

Exercise 6

1. You can go to the Control Panel, and click Add or Remove Programs, you can uninstall any application by clicking on it and delete.
2. You can log onto the website and find the SQL Connector download page. There you can get an ODBC driver for SQL Server. Install the Windows version, and you will be able to use Access to retrieve data from SQL databases.

Exercise 7

Sample answer:

From: qiangqian@sohumail.com
To: customerservice@hitech.com **CC:** **BCC:** **Subject:** complaint about the recently bought hardware **Date:** May 10th, 2009
Dear Sir/Madam, The hardware I bought from your computer shop last Friday, May 1st, 2009 was terribly flawed. It didn't fit well with my computer. Please send someone to have a look at it for me. I will wait for your man. Yours, Qian Qian Office director

Unit 8 With Customers

Exercise 1

1. F   2. F   3. F   4. F   5. T

Exercise 2

1. T   2. F   3. T   4. T

Exercise 3 (*Possible answers*)

1. product-laptop
2. We would like to ask for an exchange.
3. this weekend

Exercise 4

1. A   2. B   3. C   4. A

Exercise 5

1. E   2. D   3. G   4. A   5. B

Exercise 6

1.
  (1) Mr. Jack
  (2) forgot to call you back
  (3) all the time
  (4) the inconvenience

2.
  (1) Can I help you
  (2) convenient for you
  (3) Sorry
  (4) is full
  (5) be all right
  (6) welcome

Exercise 7

Answer the questions by yourself.

Exercise 8

Sample answer:

From: customerservice@hitech.com
To: qiangqian@sohumail.com
CC:
BCC:
Subject: Re: complaint about the recently bought hardware
Date: May 12th, 2009
Dear Mr. Qian, I was distressed to learn that the hardware you bought from our computer shop last Friday, May 1st, 2009 was not in good condition. It was due to bad shipping condition when it was being delivered to your office. Our repair person will get to fix it for you this afternoon. We will replace it if it is necessary. Please wait for him. Best wishes! Yours, Liang Daxin After Sales Manager

Unit 9 Solutions

Exercise 1

1. T   2. T   3. F   4. F   5. T

Exercise 2

1. Content templates.
2. Design templates are predesigned formats and complementary color schemes with preselected background images you can apply to any content material (the outline) to give your slides a professional, customized appearance.

Appendix B   Key to Exercises

Exercise 3

1. A   2. A   3. A

Exercise 4

1. B   2. C   3. A   4. B   5. A

Exercise 5

1. Creative Software.
2. Spreadsheets.
3. Data of accounting department.

Exercise 6

Answer the questions by yourself.

Exercise 7

To: Ma Gang		**From:** DreamTech
Fax: 5536-2588		**Page:** 1
Phone: 6636-2258		**Date:** 05/06/2009
☐ **Urgent**	☐ **For Review**	☐ **Please Comment**
☐ **Please Reply**		☐ **Please Recycle**

Dear Mr. Ma

This fax is to inform you that the third chain store of DreamTech is now open to meet our customers.

Our new store offers the same range of computer software packages as our existing stores in the new area.

Enclosed is a list of the items which are available.

Please come and visit it.

Yours truly
Yifan Lin
Sales Manager

203

Unit 10   Computer Security

Exercise 1

1. A computer virus is a computer program that can copy itself and infect a computer.
2. There are two types of viruses, benign and malignant.
3. The removable media that can carry computer viruses can be a floppy disk, CD, USB drive, or by the Internet.
4. A computer virus is a computer program that can copy itself and infect a computer. A worm, however, can spread itself to other computers without needing to be transferred as part of a host.

Exercise 2

1. D   2. E   3. A   4. B   5. C   6. F

Exercise 3

1. Kevin Minick.
2. Nicholas Whitely is arrested in connection with virus propagation.
3. Fifteen-years-old.
4. Kevin Poulsen, He is accused of theft of US national secrets and faces up to 10 years in jail.
5. German Chaos Computer Club.

Exercise 4

1. The question of security is crucial when sending confidential information such as credit card numbers.
2. Netscape Communicator and Internet Explorer display a lock when the web page is secure and allow you to disable or delete "cookies".
3. SET (secure electronic transations).
4. The only way to protect a message is to put it in as sort of "envelope", that is, to encode it with some form of encryption.
5. The most common methods of protection are passwords for access control, encryption and decryption systems and firewalls.
6. Viruses can enter a PC through files from disks, the Internet or bulletin board systems.
7. A <u>firewall</u> is hardware and/or software that protects computers from intruders.
8. <u>Sniffer</u> software prevents your being tracked while you're surfing.

9. The cookies tracks consumer fraud of all kinds.
10. Your password is the key to most information about you.

Exercise 5

Answer the questions by yourself.

Exercise 6

(1) Computer   (2) performance   (3) 4MB   (4) video game   (5) notebook

Unit 11　The Development Environment

Exercise 1

1. No.
2. Low-level language is similar to the machine code version; high-level language is similar to human language.
3. An assembler is a special program that can converted the program into machine code.
4. Compiler is a special program that can converted the higher-level languages into lower-level languages.
5. Languages, such as BASIC, COBOL, FORTRAN or Pascal, are known as source programs. Higher-level languages are converted into a lower-level language by means of a compiler, which generating the object programs.

Exercise 2

1. Java was invented by Sun Microsystems.
2. Java was developed in 1995.
3. Yes, Java is a Cross-platform language.
4. Small Java Programs, called "applets", let you watch animated characters, play music and interact with information.

Exercise 3

1. C   2. C   3. A

Exercise 4

1. Answer the questions by yourself.
2. Answer the questions by yourself.

3. Answer the questions by yourself.

Exercise 5

1. Answer the questions by yourself.
2. Windows XP does not provide Java Running Environment, you can visit the homepage of Sun Microsystems and download a Java Virtual Machine first and install it in your system. The Scheduler Home Edition will run on the new XP system tasily.

Exercise 6

CV

> # CV
> ………..(your name)……………
>
> **Personal information** (name, address, E-mail, contact numbers, etc. don't write too much)
> (Your current/permanent address)
> (Your contact number/mobile number)
> (Your E-mail address)
>
> **Career target or job objective** (list the positions you want to apply for to show your expectations)
> (Write down the position you want to apply for.)
>
> **Profile** (brief your abilities, strengths, etc. use expressive adjectives + noun phrases)
> (Introduce yourself concisely, positively and highlight your abilities and good points.)
>
> **Education** (include the time, the degree or diploma, the major and the school, relevant coursework, etc.)
> (From what time to what time, took what course, achieved what qualifications in which university and attach a school record or give details of subjects you learnt.)
>
> **Work Experience** (write the time, job title, the name of employer, duties/responsibilities, etc.)
> (If you don't have any work experience, write any social work you did in your free time or during your holidays. Mention the job title, the employer and your duties.)

Other skills (computer skills and foreign language skills)
(Offer the skills that will support your objective position, especially computer skills if you apply for a job related to it.)

Qualifications/Certifications (list all degrees, diplomas, certificates)
(List all your qualifications again clearly.)

Hobbies (mention the good ones which help enhance your image)
(Write the ones which make you look good.)

Letter

June 20th, 2014

Dear Sir/Madam,

Further to your job advertisement in today's "Harbin Evening Times", I am writing to apply for the post of the computer programmer with your company. The qualifications I have perfectly meet your needs.

I studied Java Programming which you required for the position. I am also good at working with a team to complete individual tasks appropriately, punctually and responsibly.

Please find the enclosed CV.

Your company attracts me most is that you use Java to program software. Also I prefer to work for a new creative and active company rather than a quiet, conservative corporation.

I am now preparing for my first job. Therefore, I am able to start working in June, 2014.

Moreover, I am available for the job interview at any time which suits you better.

Thanks for your careful consideration to my application. I look forward to hearing from you.

Yours faithfully,

…………………

Unit 12 New Technology

Exercise 1

1. T   2. F   3. F   4. T

Exercise 2

1. touch-screen
2. sound or light waves
3. pressure from your finger
4. iPhone's

Exercise 3

Answer the questions by yourself.

Exercise 4

Sample answer

Experience Certificate

To whom it may concern,

This is to certify that Mr. Gao Wenjun was working at Farview Information & Technology as computer system analyst from Aug 2006 to June 2008.

During his time with us, Mr. Gao Wenjun, has been a dedicated and valuable employee and he has worked hard at all tasks I have given him. He is self-confident and is a real professional. He has always exhibited reasonable judgment and he is a trusted worker. He is quick to take initiative and I am very satisfied with his performance. He has been quite helpful in the advancement of our company.

Moreover, I would like to emphasize that he can efficiently work in a team. All my staff is pleased with him and feels comfortable in cooperating with him for the realization of organizational goals and objectives.

We wish him all the best in his future career.

Yours truly,

Key to Grammar Exercises

Unit 1 Tenses 时态

1. provides
2. did; make
3. will give
4. was; have developed
5. will be studying
6. will have finished
7. was developed
8. had passed
9. is conducted
10. was leaving

Unit 2 Passive Voice 被动语态

1. will be designed
2. was discovered
3. was dismissed
4. have been established
5. were being unloaded
6. were taken
7. was discovered
8. will have been completed

Unit 3 Sentences 句子

I. 1. Positive sentence
2. Negative sentence
3. Yes-No question
4. Wh-question
5. tag question

6. Imperative sentence
7. Exclamatory sentence

II. 1. Subject clauses
2. Object clauses
3. Predicative clauses
4. Defining relative clauses
5. Non-defining relative clauses
6. Adverbial clause of place
7. Adverbial clause of time/Time clauses
8. Adverbial clause of condition/Conditionals
9. Adverbial clause of manner
10. Adverbial clause of degree

Unit 4   Nominal Clauses 名词性从句

| 1. that | 2. What | 3. if/whether | 4. Which |
| 5. that | 6. how | 7. When | 8. Where |

Unit 5   Reported Speech 间接引语

1. Diane said that Polly worked hard.
2. Diane said that she was going to buy a laptop.
3. Diane asked why they didn't go with them.
4. Diane asked to leave her alone.
5. Diane asked if she was a Star College student.

Unit 6   Adverbial Clauses 状语从句

| 1. since | 2. Although | 3. Wherever | 4. When | 5. as if |
| 6. as | 7. so that | 8. because | 9. until | 10. that |

Unit 7   Relative/Attributive Clauses 定语从句

1. that	2. which/that	3. which/that	4. whose
5. where	6. when	7. why	8. Whoever

Unit 8   Solutions Relative/Attributive Clauses and Appositive Clauses 定语从句和同位语从句

1. appositive: a computer teacher
2. adjective clause: who is a supervisor in his department
3. adjective clause: who is the principal's daughter
4. appositive: the principal's daughter
5. appositive: the head of Computer Science Department
6. appositive: the famous film star
7. adjective clause: who is a professional photographer
8. appositive: a good country woman

Unit 9   Modal Verbs 情态动词

1. can	2. be able to	3. could	4. must
5. mustn't	6. don't have to/needn't	7. must	8. should
9. shouldn't	10. May	11. must	12. can't

Unit 10   Inversion 倒装

1. Were	2. Had	3. Little	4. Never	5. then
6. Here	7. Rarely/Seldom	8. Not	9. when	10. as
11. Gone	12. Up	13. better	14. Aren't	15. May

Key to Grammar Exercises

Unit 11 Non-finite Verbs 非限定动词

1. selling   2. having checked   3. falling
4. Surprised 5. to say   6. do

Unit 12 Infinitives&Gerunds 不定式&动名词

1. swimming   2. to swim   3. crying/cry   4. meeting/having met

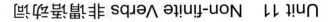